THE
RISE AND FALL
OF
NUCLEARISM

Sheldon Ungar

THE RISE AND FALL OF NUCLEARISM

*Fear and Faith as
Determinants of the Arms Race*

*The Pennsylvania State University Press
University Park, Pennsylvania*

Library of Congress Cataloging-in-Publication Data

Ungar, Sheldon.
    The rise and fall of nuclearism : fear and faith as determinants
of the arms race / Sheldon Ungar.
          p.   cm.
    Includes bibliographical references and index.
    ISBN 0-271-00840-7 (cloth : acid-free paper) —ISBN
0-271-00841-5 (paper)
        1. Nuclear weapons—Psychological aspects.   2. Nuclear energy—
Psychological aspects.   3. Arms race—Psychological aspects.
I. Title.
U264.U53   1992
355.02'17'019—dc20   91–36222
                                                                          CIP

It is the policy of The Pennsylvania State University Press to use
acid-free paper for the first printing of all clothbound books.
Publications on uncoated stock satisfy the minimum requirements of
American National Standard for Information Sciences—Permanence of
Paper for Printed Library Materials, ANSI Z39.48–1984.

I dedicate this book to my sons, Joshua and Darcy, who are lucky enough to have been born at a time when the weight of nuclear fear has been significantly reduced.

Contents

Acknowledgments

I would like to thank John Lee, Malcolm McKinnon, Arthur Rubinoff, Julian Tanner, and Morton Weinfeld for their informative feedback on all or part of the manuscript. Several anonymous reviewers also made significant contributions to the work. A special thanks should be given to the staff at the Bladen Library, most particularly Elizabeth Seres, for their helpful support. Recognition should also go to Frank Austin for his editorial assistance in the preparation of the book. Finally, I would like to thank my wife, Bev, for affording me the opportunity to write the book and encouraging me in the effort. She also scrutinized the manuscript to make sure that it remained clear and readable. For all their efforts, I remain fully responsible for the text as it appears.

Introduction

Nuclear developments during the Ronald Reagan presidency ought to be regarded with utter incredulity. His eight years in office began with the largest nuclear buildup in history and a near doubling of the U.S. military budget. But his miscalculations and miscues, particularly the loose talk about prevailing in a nuclear war, led to the resurgence of the peace movement and the narrow defeat of a nuclear-freeze motion in Congress. This opposition prompted him to announce the Strategic Defense Initiative ("Star Wars") prematurely and in terms that were unmistakably religious.[1] If Star Wars was unexpected, it was followed by a truly historic reversal. The president, who had long regarded arms-control talks as forums for Soviet propoganda, reached an agreement with the USSR to remove all intermediate-

---

1. On religious themes in the arms race, see Chernus (1986), Benford and Kurtz (1987), and Lifton and Falk (1982).

range nuclear forces from Europe. For the first time, the superpowers were about to reduce their stockpiles rather than merely put limits on their growth. There was even the hope of a 50 percent reduction in strategic arms. This "outbreak of peace" undermined the thrust of the peace movement and caused the public culture to retreat, once again, into the sanctuary of "nuclear forgetting."[2]

Little of this was anticipated, and little of it can be explained by conventional theories of the arms race. Yet such rapid shifts in sentiment are common in the nuclear age. Indeed, we will see that the public culture surrounding the bomb can be characterized by relatively distinct periods of "panic and reaction" on the one hand and passivity on the other.[3]

These radical shifts are reproduced as psychological ambivalence. Consider the paradox raised by Robert Lifton, whose work on the psychology of nuclear weapons is unequaled. He defines nuclearism as "the embrace of the bomb as a new 'fundamental,' as a source of 'salvation' and a way of restoring our lost sense of immortality."[4] Yet the most significant concept that he has added to nuclear language is that of psychic numbing. Numbing involves the use of psychoanalytic defense mechanisms to block images and feelings associated with fear of the bomb.[5]

Can we worship that which we seek to forget? Here I argue yes. Indeed, I take paradox and ambivalence as inherent in the nuclear world and as the essential starting point for coming to terms with it. More specifically, the major contention of this book is that the nuclear world is characterized by a dynamic interaction between nuclearism, the faith in this awe-inspiring power as a source of salvation, and nuclear fear, the explosive panics unleashed by this power.

The claim that the nuclear world is characterized by this dynamic interaction is consistent with the idea that the bomb was received as a *numinous* entity. While a fuller definition of numen is developed in Chapter 2, it can be understood provisionally as the feelings (rather than the ideas or rational attributes) associated with religious phe- nomena. These feelings are of a twofold nature: at one pole, a sense of awe and overwhelming power; at the other, a sense of demonic dread and overwhelming fear.[6] Consistent with this formulation, the

2. The term is used by Gamson (1987).
3. On panic and reaction, see McDougall (1985, 422) and Ungar (1990a). On passivity, see Boyer (1980).
4. Lifton and Falk (1982, 87).
5. Lifton and Falk (1982, 103–4).
6. The best source on the numinous remains Otto (1958). See also Chernus (1986).

bomb was received as a source of transcendental power that was expected to be decisive in the political, military, diplomatic, and economic realms. At the same time, it was the source of demonic fear that conjured up images of vaporized cities.[7] Here I will argue that it is the ongoing interaction between the two poles of this confounding power that accounts for many of the paradoxes and radical shifts that characterize the nuclear world.

Since the power and the fear were present from the bomb's inception, it was not inevitable that one or the other would predominate. However, I contend that events in the first half of this century made it highly likely that the sense of power initially prevailed. Histories of the bomb typically focus on its immediate antecedents: the scientific breakthroughs that coalesced with the dramatic race to beat a Nazi bomb. These histories are primarily concerned with the question of how and why the bomb was built when it was. Here I broaden that question and ask how historical factors influenced the *reception* of the bomb in the West. In other words, an understanding of nuclear faith necessitates placing it in the historical and religious context that preceded it.[8] Where historians of the nuclear age invariably begin their tale with the revolution in physics, I recommend the value of looking back further in time and concentrating on quite different events. The goal is not only to retell the nuclear story but to do so in such a way as to preserve and cultivate an attitude of utter incredulity.

The historical circumstances that led the West to invest such faith in the bomb are so vast that they are best grasped by metaphor. Here I draw on Nietzsche's idea of the Great Collapse, which encapsulates his concern with the "death of God" and the manifold consequences of the loss of faith for Western societies. Following Nietzsche, I define the Great Collapse inclusively. For our purposes, the Great Collapse is constituted by the concatenating historical events—including World War I, the Bolshevik Revolution, the rise of fascism, and the Great Depression—that not only threatened the basic values but at times the very survival of the Western democracies. In threatening values, the Great Collapse brought to a culmination the progressive "disgodding" of social reality that had begun centuries earlier. Thus it was in these years that the faith in the inevitability of progress was undermined. Progress, as understood in the nineteenth century, was based

7. On overwhelming power, see Herken (1980), who terms the bomb a "winning weapon." The best discussion of fear is Weart (1988). See also Blackett (1949). Boyer (1985) is one of the few writers to grasp both ideas simultaneously.

8. See, for example, L. Lamont (1965); Clark (1980); and Wyden (1985). The most recent study by Rhodes (1986) does consider the prior history in more detail.

on power *and* morality; and, in most instances, the faith in progress was upheld by a distant Providence. But the Great Collapse trampled the faith in moral progress and the belief in a providential history.[9]

Many of the same forces that sundered the faith in progress posed imminent threats to the besieged bastions of democracy and capitalism. Yet the democracies found themselves unable to respond collectively. The absence of an effective balance of power and collective economic solutions resulted in drift and paralysis. Most important, the sociopolitical factors that undermined balancing in the interwar years either carried through the war or appeared likely to reemerge as World War II ended. It is in this context of persistent threat and the feared absence of an effective postwar balancing system that Western political leaders invested their faith in a technological panacea, namely, the bomb.

The unleashing of such a transcendent power in the context of the Great Collapse resulted in an excess of positive symbolic and practical meanings being attributed to nuclear armaments. As a "winning weapon," the bomb was construed as far more than an increase in firepower: not only was it expected to be decisive in the political, diplomatic, economic, and military realms but it also came to be associated with elements of American civil religion. In their totality, the positive meanings conferred on the bomb constitute what Lifton terms nuclearism. He defines nuclearism as "a secular religion, a total ideology in which . . . the mastery of death and evil . . . are achieved through the power of a new technological deity." He suggests that the "nuclear believer or 'nuclearist' allies himself with that power and feels compelled to expound on the virtues of the deity. He may come to depend on the weapons to keep the world going."[10]

The idea of a bomb that harnessed the fundamental power of the universe gave rise, above all, to a sense of omnipotence. With total power, human domination and control—the ability to subdue evil and to transform nature—are deified. In a world where God's *social* reality had been rendered superfluous, humanity, or at least a part of it, could play Providence for itself. Omnipotence, in short, is the dynamic quest for and faith in transcendent power.[11] Such power is more than normal; it is supernatural, the power that properly belongs to immortals, not mortals. And here was a power that was both

9. On this period, see Hobsbawm (1987).

10. Lifton (1979, 369). Unlike Lifton, who stresses the secular nature of nuclearism, I contend here that the bomb revives primitive aspects of religion. Weart (1988) is an excellent source for atavistic associations with the nuclear world.

11. On the idea of omnipotence, see Chernus (1986) and Wolfe (1984).

metaphorically and practically transcendent. Metaphorically, it represented a sense of power that had hitherto been limited to the gods. Splitting the atom dramatically heightened the sense of human dominion; it practically elevated us into the empyrean. The control over nature's ultimate power was also taken as a sign of grace, an indication of America's moral superiority and redemptive capacity.

Practically, transcendent power embodied one of the most critical attributes of a deity: dominion over life and death, the power to preserve or to deny the future. The bomb appeared as a godsend. It seemed to be the answer to the Great Collapse. Not only could it remove the threat posed by totalitarianism, but it seemed capable of securing the future, of assuring historical continuity to Western democracies and values. Finally, it afforded the promise of a utopian future, one based on (nothing more than) cheap and abundant energy.

However, the bomb was no ordinary technological wonder, not simply another accretion to the pantheon of technological progress touted in world fairs and exhibitions. Nuclear power was "wholly other." As one theologian put it, the bomb was "out of sync" with the powers and responsibilities conferred on humanity by Christianity.[12] Thus omnipotence is only half the story, reflecting the "power" pole of the numen. Allied to it and never far below the surface is the sense of nuclear fear. The dark side of the bomb was visible at its inception. The emerging capacity to extinguish all life—itself a key attribute of a deity—creates, in Lifton's terms, "an insistent awareness of absurdity." The resultant ambivalence surrounding the bomb is in fact characteristic of primitive deities. In a coincidence of opposites, such gods combine good and evil, awe and terror, beneficence and hostility.[13]

The analysis of nuclear fear and its interactions with the sense of omnipotence will be a central consideration here. Overall, I contend that people prefer to avoid confronting the nuclear threat and tend to respond to it only when it is forcefully impressed on them. Thus I argue that there is a generalized and relatively low level of nuclear fear that, on occasion, is punctuated by *moral panics*.[14] Significantly, these panics can be caused by either American or Soviet nuclear developments. The most extreme panics, however, are unleashed by the perception of spectacular and startling Soviet challenges to American nuclear hegemony. Since these episodes greatly exaggerate So-

12. From Kaufman (1985, ix).
13. On primitive deities, see Russell (1977).
14. On moral panics, see Cohen (1972) and Ungar (1990a).

viet challenges, they give rise to moral crusades that result in massive nuclear buildups and other efforts to restore the American sense of nuclear omnipotence.

Here I arrive at the crux of my thesis. While military, bureaucratic, technical, political, and economic factors undoubtedly had important impacts on the development of the bomb and the arms race, I argue that the quest for transcendent power and the moral panics engendered by Soviet threats to that power also had independent and highly significant effects on these developments. The dynamics of the American effort to preserve a sense of omnipotence, embodied in a "panic and reaction" syndrome, impart an impetus to the arms race over and above that which can be ascribed to any or all of the factors listed above.

The working out of the bomb's numinous aspects—the relationship between nuclearism, the faith in this awe-inspiring power as a source of salvation, and nuclear dread, the extraordinary fear that can be unleashed by this numinous phenomenon—is a central and ongoing dynamic of the nuclear world. At different times and among different publics, the bomb is quite capable of engendering radically different responses. Massive nuclear buildups, millions of protesters trying to block deployments, and passivity or nuclear forgetting: these are among the possible responses to nuclear realities. Here I apply the two themes of omnipotence and fear to the development of the bomb and to the ensuing arms race and seek to show that the paradoxes and rapid shifts that characterize the nuclear world are strongly related to its numinous foundations. Historically, I seek to trace the shifting balance between the fear and the faith. Although the faith came first, and the two coexisted in an unhappy tension for much of the nuclear era, ultimately it is the fear that has come to predominate and now defines the nuclear future.

In assessing the Soviet response to the bombing of Hiroshima, D. F. Fleming notes that they "were bound to be deeply influenced by the entire history of their relations with the West since 1917, while the West would think only of the future."[15] Contrary to Fleming, there are no good grounds for exempting the West from the influence of its own history. The historical principle that he applies to the Soviets can and should be applied to the reception of the bomb in the West. The bomb was a new factor, but the meanings conferred on it and the plans developed for it were significantly influenced by events since at least 1914.

15. Fleming (1961, 325–26).

Chapter 1 outlines the historical context that made faith in nuclear omnipotence conceivable. It shows how the idea of omnipotence derived from religious and scientific trends in the West. Specifically, the "process of disenchantment" led to the inadvertent and lingering death of God. In lieu of God, public faith was transferred first to reason and then to progress. But the faith in progress was shattered, symbolically and practically, by the Great Collapse. The concatenating threats and disasters that constituted the Great Collapse stripped away all limits and moral concerns from the idea of progress and left a problematic legacy: sheer human mastery of the natural world.

Chapter 2 shows how the idea of omnipotence was beginning to form around the scientific enterprise and was then put on the map by the atomic bomb. The sense of omnipotence that arose from the bomb entailed a qualitative leap in human mastery: the scientists had possession of godlike powers, powers that were unhuman in their scope and implication. The bomb was, in effect, a numinous phenomenon, combining extraordinary power and extraordinary fear. While the scientists were originally motivated by the quest for godlike powers, the associated sense of fear became palpable in the aftermath of the Trinity explosion.

Chapter 3 examines how (most) politicians immediately ascribed a sense of omnipotence to the bomb. Given the persistent threats to the democracies and the failure of collective action, politicians were primed to invest their faith in a technological panacea. Thus the bomb was construed as a winning weapon—but unlike the case of previous weapons, faith in the bomb far transcended its possible military uses: it was expected to be decisive in the political, economic, diplomatic, *and* military realms. In this regard, the bombing of Hiroshima and Nagasaki can be regarded as efforts to test and to display this new power.

Chapter 4 examines the public reception of the bomb. Here the process is reversed, with the fear preceding the sense of power. Hence there was a pressing need to defuse the fear and to confer a sense of legitimacy on a power that seemed to transcend the limits of human control. Several means were used to tame the fear, the most significant being the alignment of the bomb with elements of American civil religion. Thus it was construed as either a direct bequest of the deity or as a consequence of the special destiny of the American nation. But the linkage between the bomb and the historic destiny of the nation had a number of striking consequences: it put the arms race at the center of the international conflict over which of the two rival systems would predominate historically; it rendered the panics unleashed by Soviet nuclear advances acute, for these gains threatened not only national security but the most sacred beliefs about the country's destiny; and it fostered an ongoing search for the definitive technological solution to the Soviet threat. The chapter ends by discussing the role of the bomb in the start of the Cold War.

Chapters 5 through 7 examine the moral panics unleashed by Soviet challenges to the American sense of nuclear omnipotence. They deal, respectively, with the Soviet atomic bomb–Korean War panic, the *Sputnik* panic, and the Cuban missile crisis. Each episode is discussed by analyzing the striking Soviet nuclear advance that stoked the latent nuclear fear (and threatened national security and the sanctity of the American way of life) and the consequences of the panic for arms buildups and American efforts to find the definitive solution for the nuclear threat. The analysis means to demonstrate that major nuclear buildups are *best* understood as panic-induced phenomena. It also seeks to reverse the conventional (and often uncritical) wisdom concerning the military-industrial complex (MIC). Specifically, it intends to show that the panics were largely authentic outbursts of fear that the MIC did not create. Rather, the much-vaunted power of the MIC functions primarily to impart direction to the panics. Finally, the analysis seeks to make plain that the scope of the panics cannot be understood unless we consider their symbolic aspects. In other words, the panics were so extreme because they simultaneously evoked nuclear fear *and* challenged sacred beliefs about the special destiny of the nation.

Chapter 8 examines the radical reversal that followed the *Sputnik* and missile-crisis panics. The public culture of the time is character- ized by "nuclear forgetting," a concerted effort to suppress the cascading fears and obsessions that had gripped Americans for two decades. Nuclear forgetting was integral to détente and to the signing of the Partial Test Ban Treaty of 1963. It also served to attenuate and tarnish the faith in nuclearism. Paradoxically, however, nuclear for- getting did not slow the pace of the arms race. Rather, the missile crisis sanctified the view that the Soviets could not be trusted. They were bearers of "totalitarian omnipotence," a metaphor that uncriti- cally merges the Soviets with Satan. The interactions among nuclear forgetting, totalitarian omnipotence, and nuclearism are analyzed in the post–missile crisis world.

Chapter 9 considers the ultimate consequence of totalitarian om- nipotence: the abiding fear of a Soviet first strike. The chapter analyzes the final moral panic produced by the confluence of the "window of vulnerability," the Iranian crisis, and the Soviet invasion of Afghanistan. Unlike previous panics, this one was created from above and culminated in a startling reversal: a genuine "war-fighting" panic produced by the loose statements and other actions of the Reagan administration. The chapter argues that this war-fighting panic crushed whatever residual faith remained in nuclearism.

The Conclusion examines the implications of the bomb's confounding power in the new world authorized by the Gorbachev revolution. It suggests that while Gorbachev was able to lighten the burden of nuclear fear, that factor still remains far more significant than the shattered faith in nuclearism. While there appears to be nothing on the horizon that could revive the faith (including the "smart weapons" used in the Gulf War or the coming generation of "brilliant weapons"), nuclear fear could again be a significant factor, this time in the context of Third World proliferation.

1

God, Progress, and the Great Collapse

Writing in 1692, T. Doolittle reports that in the London earthquake of that year no one died and no house was destroyed. The earthquake did shake the roofs and foundations of houses. It also shook the prisons, where the locks opened and prisoners escaped. Doolittle concludes that the event could only have one meaning: it was clearly intended as a warning, a portent of greater calamities to come, unless people learned to forsake sin and turned to God. London was no Sodom. Yet it had been subject to a number of unusual premonitory signs:

> Plague, Fire and Earthquakes are amongst the most terrible Judgments, whereby the great and holy God doth manifest his displeasure against a sinful People, and his indignation and hatred of their heinous and crying Sins. *These three* have befallen *London*. The *First* in the GREAT DYING YEAR, 1665, the

Second in the GREAT BURNING YEAR, 1666, and the *Last* in the GREAT TREMBLING PRESENT YEAR, 1692.[1]

Doolittle is not simply exhorting the faithful to ever-greater acts of piety and contrition. Rather, he is engaged in a polemic against those who speak of the alleged "natural causes" of earthquakes. He allows for the existence of such causes, but severely limits their efficacy. Earthquakes that issue from them are not simply the result of "chance"; they are woven into the broader design of the Creator. But it is far more important that earthquakes can arise from the immediate hand of God without the presence of any natural causes. London is decidedly an instance of the latter. For how could something so powerful be so restrained in its effects?

Could all the mighty *Captains, Lord Generals, Kings* and *Emperours,* with all their armies . . . cause such a shaking of the Earth as that *so lately* was? . . . If men *could shake* the earth and your Houses as God did, could they so securely have *preserved* them from falling down, and you from hurt, as God did?[2]

Doolittle's musings contain the germ of two ideas that concern us here. The first is the process of disenchantment, in which the Christian cosmology is challenged and eventually replaced by the scientific, or naturalistic, outlook.[3] According to Herbert Butterfield, the historian of science, the process of disenchantment, five hundred years in the making, is unrivaled by any historical event after the rise of Christianity: "it changed the character of men's habitual mental operations . . . transforming the whole diagram of the physical universe and the very texture of human life itself."[4]

The second theme is the divinization of human powers and history. That is, the process of disenchantment *necessitated* the exercise of human powers. Without Providence, the destruction of a city could be equated to the crushing of an anthill by a roaming child; people were simply vulnerable to the whims of nature. In the long run, the emergent objective would be to defy fate and exercise control over chance itself. To Doolittle's once-rhetorical question—could men

1. Doolittle (1692, A1–A2); italics and emphasis in original. See also Kendrick (1956).
2. Doolittle (1692, 23–25); italics in original.
3. The phrase is used by Weber (1946, 129).
4. Butterfield (1957, vii–viii).

cause such a shaking of the earth—we can now answer in the affirmative. Nuclear weapons afford us such a power, if we choose to exercise it.[5] We can even answer yes, theoretically, to Doolittle's second question. Neutron bombs and limited nuclear war are intended to shake the earth and still preserve something of value. Effectively, then, power and meaning were to become entwined. As God was pushed away from His creation, we increasingly sought to recoup His lost powers. These two themes—disenchantment and the divinization of humanity and history—provide the background for comprehending the reception of the bomb and the emergence of nuclearism.

From Moving Spirit to Moving Force

The process of disenchantment involves the steady and cumulative "disgodding" of the physical world and social reality.[6] For our common understanding of the world, as embodied in the naturalistic outlook, God barely exists. Where it is not too much to claim that God was the main actor in the Christian cosmology, He hardly acts at all any more. Although some references are still made to Him in the public culture, these are, as will be seen in the discussion of the American civil religion, essentially perfunctory.

Disenchantment had one particularly historical consequence that is of interest to us: specifically, it led to the improvisation of new faiths in which humanity aimed to play Providence over itself. Here I outline the process of disenchantment in the first scientific revolution and seek to show that scientific presuppositions not only eroded God's physical and social reality but underwrote the expanding confidence that made faith in reason, progress, and ultimately omnipotence possible. More specifically, this chapter traces the movement from the

5. A geographer told me that the atomic bomb dropped on Hiroshima was far less powerful than a small earthquake. However, there is no theoretical limit to the size of hydrogen bombs. When the Soviets exploded a bomb in the fifty-megaton range, the size of the blast apparently frightened even them. The weapons have become smaller but more accurate. What remains significant, however, is how willing we are to compare our powers with those of nature and of God.

6. As defined here, the death of God entails the loss of His *public* reality and is not concerned with the standing of either Christianity or personal belief. A view consistent with the one presented here can be found in Turner (1985). Peter Berger (1969) develops the idea of God losing His historical and institutional relevance while retaining (a diminished!) personal relevance. Disenchantment proceeds by removing not only God but all magical and supernatural elements from the common worldview.

scientific disgodding of reality to the belief in reason and then in progress. It goes on to develop the idea of the Great Collapse, which both undermined the faith in progress and put Western democracies at acute risk. The upshot of these menacing developments was to create a historical context that not only provided a compelling stimulus to build the bomb but also assured that it would be received as a winning weapon that embodied the idea of omnipotence.

The effects of the scientific revolution on the Christian cosmology were slow, often imperceptible, and, above all, inadvertent. Renaissance naturalism was still struggling to differentiate between occult causes and the mechanisms of nature, regarded as a self-sufficient system under the guidance of natural law. The boundaries between law and mechanisms on the one hand, and magic and miracles on the other, were fuzzy and shifting. What we now regard as sheer superstition (e.g., the idea of animals prophesying) was then regarded as part of the natural working of nature.

The struggle to sort out these convoluted processes is apparent in the work of the German astronomer Johannes Kepler. His first book, the *Mysterium Cosmographicum,* published in 1596 when he was only twenty-five years old, is divided into two parts that correspond with the split consciousness of an age of transformation. In the first section, which is mystical, Kepler seeks to uncover the numerical harmonies governing nature. He discusses the five "perfect spheres" and then notes that there are five intervals between the six planets. This symmetry reflected a divine arrangement. It explained not only why there were just six planets, but also accounted for the distances between them, which were large enough so that each interval could accommodate one of the perfect spheres.

Then we come to a metamorphosis:

> What we have so far said served merely to support our thesis by arguments of probability. Now we shall proceed to the astronomical determination of the orbits and to geometrical considerations. If these do not confirm the thesis, then all our previous efforts have doubtless been in vain.[7]

Kepler, a man steeped in astrology and metaphysics, has suddenly, without warning or apparent rehearsal, vaulted into a different world. The teleological arguments spelling out God's intent are reduced to "probabilities." Truth is a function of observations, of facts.

7. Cited in Koestler (1959, 255).

This transformation from metaphysics to scientific empiricism took considerable time to be worked out in Kepler's mind. The first volume of the *Mysterium* offers two possible explanations for the rate of movement of the planets: either there are separate souls moving each planet or there is a single soul in the center of all orbits. The second volume, revised almost twenty-five years later, contains the following changes:

> (ii). That such souls do not exist I have proved in my *Astronomia Nova*.
> (iii). If we substitute for the word "soul" the word "force" then we get just the principle which underlies my physics of the skies in the *Astronomia Nova*. . . .[8]

This is a remarkable step in the hesitant emergence of the modern outlook. The shift in fundamental concepts from animism to naturalism is a decisive stage in the paradigm shift that is under way. The divine living machine is giving way to the divine clockwork. Final causes are giving way to the study of efficient, proximate ones; "why" is becoming "how." God is removed from His place "above" to somewhere "out there." His intervention in the world is slowly but firmly reduced, until it is no longer considered necessary.

While others made God walk the proverbial plank, it was Pierre Laplace, the French astronomer and mathematician, who sent Him to the bottom. Laplace showed that the solar system was a self-regulating mechanism. In 1812 he carried the scientific hubris that would culminate in the hydrogen bomb to its extreme by advancing the notion of the Divine Calculator. Here he conjured up a mathematical physicist, who, if he knew the positions and velocities of all particles in the world at a particular moment, and also had a powerful enough calculating machine, would predict *all* the future and know *all* the past. What we have, in Laplace's fantasy, is a totally deterministic universe ogled by a consummate scientist and calculator, one who knows all the laws and relations out of which the universe is composed.

One cannot help but be amused today by Laplace's spurious sage, since modern science first canonized uncertainty and more recently chaos. According to Dante, in *The Divine Comedy*, pride is the greatest sin and the root of all other sins. Aquinas taught that Satan's ultimate sin was the wish to be God. Pride, at its worst, is mimicry of the deity.

---

8. Cited in Koestler (1959, 259).

By this reckoning, Laplace should be found in the ninth circle of the *Inferno,* perhaps just above the three-faced Lucifer, the "emperor of the dolorous realm." For when asked by Napoleon about the place of God in this scheme of things, he replied: "Sire, I have no need of this hypothesis."

Reason and Progress

When the savants of ancient Greece found their quest for immortality denied by reason, their response was Hellenic pessimism and tragedy. The philosophes of the Enlightenment were probably the first to bear the full brunt of a naturalistic outlook in a society where God's reality was not frequently questioned. An inevitable by-product of this shift in outlook is that God was less keenly felt. The anguish of discovering that the universe is a blind mechanism without moral drives is clear in Vauvenargues: "Ah! If it were true, if men depended only on themselves, if there were no rewards for the good and punishments for crime, if all were limited to this earth, what a lamentable state!"[9]

Like sociologists of the future, the philosophes regarded God or an equivalent faith as necessary to uphold social order. God sustained a meaningful sense of the cosmos, and without Him moral collapse seemed inevitable. While they revel in humanity's newfound freedom and independence, they also realize, as Voltaire put it, that the removal of God's yoke allows man "to abandon himself to all the furies of his unchained desires . . . there being no absolute virtue or vice, he can do anything with impunity."[10] The philosophes recognized that the social world was a human construction and simultaneously held that it was necessary to conceal our authorship of that world. Hence they scurried to improvise replacement deities as the Christian metaphysic crumbled and faith in the original deity was scaled down and rendered precarious.[11]

9. Cited in Crocker (1956, 7).
10. Cited in Crocker (1956, 25).
11. Peter Gay (1954, 123), who has led the effort to debunk the overly rationalistic and optimistic view of the Enlightenment, agrees with this thesis: "as men of letters at home in a world that was losing its Christian vocation, the philosophes felt this critical loss as a deep problem and solved it by reinterpreting and transforming their civilization." They were dealing with a crisis of secularization, one that Gay terms slow, subtle, and pervasive. For the luminaries it was a "grim struggle," as they caught a glimpse of the consequences that might follow.

God's heirs take over His social and historical functions. Providence, as it fell into human hands, expressed itself historically. The
philosophes deified, in various ways, reason and nature.[12] The deifi-
cation of reason was a generalization from the scientific world, particularly from Newton, who had discovered "God's Laws" and became
the idol of a new phenomenon, the "cult of great men."[13] The idea
was to take the "geometric spirit" and transport it to all branches of
knowledge. As Claude Gilbert put it, "By obeying reason we depend
on no one but ourselves and so, in a sense, we too become as Gods."[14]

But these replacements are minor deities because of a host of
limitations, ranging from their failure to supply a complete theodicy
for death to the fact that their effectiveness can be readily undermined. Thus faith in reason, in its *specific* Enlightenment manifestation, proved to be short lived. On a theoretical plane, David Hume
showed that the belief in reason was based on superficial and unsound
assumptions.[15] Advocates of natural religion could agree on the un-
deniable evidence of God's handiwork in nature, but on little else. On
a practical plane, reason became associated with the excesses of the
French Revolution and Bonapartism. By implication, the Enlightenment bequeathed a problematic idol to the next century. That century still needed the idea of reason, but, because of its revolutionary
associations, had to overstate the blind faith of the Enlightenment in
order to distance itself from, yet still preserve, its patrimony.[16]

The naturalistic outlook reached ever-wider publics in the nineteenth century as the Enlightenment faith in reason was replaced by
the belief in progress. Intellectual progress was enlarged into the
general progress of humanity, but with important differences. Still
fearful of the anarchy that they regarded as the legacy of the
Enlightenment, nineteenth-century thinkers for the most part divested progress of revolutionary ideas and allied it instead to the

12. On the Enlightenment, see Becker (1960); Cragg (1964); Crocker (1956); Gay
(1954); Hazard (1946), (1963); Vyverberg (1958); and Wiley (1940).

13. The designs for cemeteries gave prominence to men like Newton, Descartes, and
Bacon, who were to serve as models for the living. Set against the destructive force of
death is not Christian salvation but the veneration of future generations. In these
"spaces of emulation" humans were being endowed with attributes of the divinity. See
Etlin (1984).

14. Cited in Hazard (1963, 154).

15. See Cragg (1964, 167–71).

16. Peter Gay (1954) observes that the popular distortions of the Enlightenment as
overly rationalistic and optimistic have been immune to historical revision. He notes
that even dictionaries give it erroneous definitions. Here I suggest that we have created
the Enlightenment not so much that we deserved but that we needed.

notion of steady evolution. Progress was joined to the idea of "law," a benevolent necessity beyond human control. Progress, in effect, did not totally dispense with God's tutelage.[17]

At the practical level progress was forged from energy and tempered with morality. The nineteenth century discovered that nature was a fountain of power, and that this power could be transformed into energy. It was the steam boiler, and most particularly the locomotive, that became the motor of mastery. The "iron horse" was intoxicating. Its "fiercely direct energy" and "power magnified beyond comprehension" vividly embodied the idea of progress.[18] Its attributes were those of machine civilization: the replacement of wood by iron and steel; mechanized motive power; a vast increase in geographic scope, movement, and speed; and a spirit of rationality and efficiency. The first locomotive dated from 1804. In 1830 there were twenty miles of rail; in 1854, six thousand. The enhancement of human powers was unprecedented; it verged on the magical. The spectacular advances of science and technology seemed to defy any limits; the indefinite increase of human power seemed probable.

The mood of exhilaration was caught in slogans—"the conquest of nature" and the American vision of "Manifest Destiny" being cases in point. Significantly, progress was not limited to material benefits, but came to include such social and humanistic reforms as the extension of democratic rights, the spread of education, and the abolition of slavery. And to this was added the belief that the unity of humanity was a realizable goal.

17. Still debated is the question of how much progress owes to Christian ideas. While the concept has definite roots in Christianity, it is more than a mere secularization of its ideas, as is held, for example by Nisbet (1980) and White (1962). Christianity oscillated between two poles: the Fall and degeneracy at one extreme, and the dignity of humankind at the summit of the Great Chain of Being at the other. Modern versions of progress were succored by the latter pole. But progress was not a direct bequest of Christianity. It took events like the quarrel of the Ancients and Moderns for the idea to be wrested from the muck of degeneracy. The catalyst for the shift to the confident pole was more than religious; the expansion of trade and the revolution in the mechanical arts in the Middle Ages were critical to lifting the yoke of degeneracy. Yet as Rossi (1970, xi) observes, "the philosophical culture of the seventeenth century brought on a full and mature awareness of certain ideas that had slowly been making their way at the edges of the official culture, outside the academic culture and almost always in opposition to it." Bury (1955, 73) goes even further when he argues that "it was not until men felt independent of Providence that they could organize a theory of Progress." Leiss (1972, 33) further complicates the issue by observing that the same concept may well have been filled with radically different contents.

18. The phrasing is used by John Stilgoe (1983).

The Great Collapse

The undermining of the faith in progress was a protracted process.[19] Yet so fundamental and persistent were the threats to Western democracies in the first half of this century that they are best grasped by metaphor. Here we employ Nietzsche's idea of the Great Collapse:

> Much less may one suppose that many people as yet know *what* this event [God's death] really means—and how much must collapse now that this faith has been undermined because it was built upon this faith, propped up by it, grown into it; for example, our whole European morality. This long plenitude and sequence of breakdown, destruction, ruin and cataclysm that is now impending—who could guess enough of it today to be compelled to play the teacher and advanced proclaimer of this monstrous logic of terror, the prophet of gloom and an eclipse of the sun whose like has probably not yet occurred on earth.[20]

Nietzsche foresaw that European culture was moving toward a catastrophe. His prediction that "there will be wars as have never happened on earth" proved correct, though not necessarily for the reasons he proffered.[21]

Images of ruin and terror all too readily characterize the twentieth century, as if everything is permitted now that God is dead. Starting with the Great War, which punctured the idol of progress and gave rise to the slogan "Never again such innocence," this century has seen the prevalence of calculated and logical murder, which suggests that

---

19. Nisbet (1980) claims that the belief in progress was more resilient than I argue here. Much of the disagreement is a matter of terminology. The idea of progress certainly survived, but it was divested of much of its meaning, faith, and emotional properties. It is now common to say, "Progress cannot be stopped." Statements like this are highly ambivalent; progress is compressed to material benefits—benefits that combine desirability with a sense of a curse, reflecting, for example, a wide range of possible ecological disasters. But the most resounding change is in the idea of history. To believe in progress was to believe in history. We now view history ironically, or worse. See Almond, Chodorow, and Pearce (1982).

20. Cited in Küng (1981, 369–70). See also Ungar (1990b).

21. Nietzsche (1968) explains the "great collapse" as a consequence of the "dissolution of all values." Such an idealistic explanation overlooks the dynamics of capitalism, technology, drift, and error.

history has become the theater of unbridled power.[22] Here I define the Great Collapse inclusively: it is constituted by the concatenating historical events that not only undermined the faith in progress but also threatened the basic values and, at times, the very survival of Western democracies. It was precisely this confluence of symbolic and real threats that rendered the Great Collapse so critical and created the preconditions for investing so much faith in the bomb. For an essential aspect of the cumulating threats was the failure of conventional solutions and hence the readiness to accept a technical panacea that seemed godlike in its magnitude.

Signs of the impending catastrophe manifested themselves well before the "natural break" of 1914. These "disquieting incidents and tensions" included economic disturbances from 1873 on, as well as the threat of the masses, who were demanding greater economic and political rights.[23] Despite the sense of menace, bourgeois liberalism still possessed a sense of resilience, based on its relative security *and* values:

> But if (unlike the decades after 1917) bourgeois society as a whole did not feel seriously and immediately threatened, neither had its nineteenth-century values and historic expectations been seriously undermined as yet. Civilized behavior, the rule of law and liberal institutions were still expected to continue their secular progress.[24]

The Great Collapse would undermine the secular faith and values and simultaneously confront these societies with pervasive threats, the enemies at the gates.[25] The collapse has both an ideal and a sociopolitical dimension.

In 1905 the world changed forever.[26] If the date is arbitrary, it is clear that the chain of progress that symbolically linked the past to the future was corroding and about to snap. By this time radical shifts

22. Cited in Fussell (1975, 18). See Camus (1956).
23. Churchill (1923, 18).
24. Hobsbawm (1987, 100).
25. See McNeill (1982, 379–80).
26. This statement was made by Virginia Woolf, who used 1910 as the critical juncture. H. Stuart Hughes, whose *Consciousness and Society* (1958) remains the most cogent analysis of the transformation in social thought, picks 1905 as the watershed. Arthur Koestler, born in that year, wrote, "I was born at the moment when the sun was setting on the Age of Reason" (cited in Baumer [1960, 187]). Einstein's Nobel Prize–winning paper was also published in 1905.

were under way in most scientific and literary domains. There was, according to Hughes, the "sense of a collapse of accepted standards."[27] A concern for the subjective and the irrational was replacing the positivist faith in reason. Freud opened the turmoil of the unconscious to public viewing. In physics, quantum theory and relativity canonized uncertainty and thereby removed the "fixed point" that had sustained the Newtonian universe. Where scientists recoiled from the implications of their ideas and tried to keep relativism in check, writers and artists became active participants in the circus of incoherence. Across all creative domains, there arose a new aesthetic sensibility: weakened references and significations; self-consciousness and privatization; the abolition of standard forms; an abstract, often hallucinatory reality; and a sense of alienation and malaise.

Many reasons have been advanced for this unprecedented convergence of sensibilities, ranging from the exhaustion of traditional artistic languages to the shift in the social position of the artist.[28] Just how much this convergence owes to the looming catastrophe in Europe and the associated loss of faith is difficult to say. Perhaps Picasso and Braque wrought the revolution in cubism without any corresponding great ideas.[29] In contrast, the literary culture was more self-conscious and exuded a sense of meaninglessness that echoed the demise of God and foreshadowed collapse.[30] But whatever the case, once Europe was embroiled in war the intelligentsia turned to this art and *found* in it premonitions of the disaster that had overtaken them. Thus the new sensibility was experienced as the artistic foreshadowing and expression of the collapse of all faith. The artistic malaise that preceded the war then coalesced, starting with the Dada movement in 1916, into an enduring literary tradition that is ironic and disillu-

27. Hughes (1958, 364).
28. See, for example, Lukács (1963).
29. This claim is made by Russell (1981, 101–2). But then Picasso is quoted as having said about his early work, "And this strangeness was what we wanted to make people think about because we were quite aware that our world was becoming very strange and not exactly reassuring" (cited in Gilot and Lake [1984, 72]).
30. Consider Tolstoy's (1940, 12) report on six years abroad: "Life in Europe and my acquaintance with leading and learned Europeans confirmed me yet more in the faith of striving after perfection in which I believed, for I found the same faith among them. That faith took the form it assumes with the majority of educated people of our day. It was expressed by the word 'progress.' It then appeared to me that the word meant something. I did not then understand that, being tormented (like every vital man) by the question of how it is best for me to live, in my answer, 'Live in conformity with progress,' I was like a man in a boat who when carried along by the wind and waves should reply to what for him is the chief and only question, 'whither to steer,' by saying, 'We are being carried somewhere.' "

sioned.[31] In this process, the arts lost the power to forge and sustain a coherent wisdom and mythology.

The sociopolitical events that threatened the democracies are so well known that they can be briefly listed. World War I and its immediate aftermath shattered the European system, politically, militarily, and psychologically.[32] According to Paul Fussell, the Great War became "all-encompassing, all-pervading, both internal and external at once, the essential condition of consciousness in the twentieth century."[33] The war helped trigger the Bolshevik Revolution and, as a result, the Paris Peace Conference sought to establish a *cordon sanitaire* to separate the Soviets from the rest of Europe.[34] Here even the United States was not immune. The Red Scare of 1919, directed against labor unions, prefigured the Cold War in its unfounded fear that communist insurrection was imminent in the United States.[35] In Europe, the fear of revolution was more realistic. Communist governments were established in Hungary and Bavaria, but the nascent regimes were crushed with either the aid or sanction of the democracies. The seeming slide toward communism in Italy was halted by the seizure of power by the Fascists in 1922. As Churchill put it, "Fascism was the shadow or ugly child of communism."[36]

The postwar economic and political crises were relieved by a brief recovery, spanning 1924 to 1929. But this was followed by the economic and industrial collapse of the depression, which particularly blunted an American optimism that had been sustained by its insulation from European politics.[37] The depression, coupled with the challenges posed by Nazi Germany ánd militarist Japan, put the democracies at acute risk. Between 1920 and 1939, representative

31. See, for example, Goudsblom (1980); Lester (1968); Lukács (1963); and Thielicke (1969).

32. For the effects of the war, see Fussell (1975). On the failure of the balance of power, see Ross (1983). General coverage of the period is given in Hobsbawm (1987) and Kennedy (1988a).

33. Fussell (1975, 321).

34. Keylor (1984, 89–91).

35. See Murray (1955).

36. Churchill (1948, 13).

37. According to Alfred Kazin: "Something happened in the thirties that was more than the sum of the sufferings inflicted, the billions lost, the institutions and people uprooted: it was education by shock. Panic, a panic often significantly disproportionate to the losses of those who were most afraid, became the tone of the period. . . . In the world after 1932, where everything seemed to be breaking up at once, the American had at first neither a sense of history nor the consolation of traditional values. He was oppressed by forces that were meaningless to him in operation and hence all the more humiliating in effect" (cited in Bloom [1986, 33]).

government disappeared from most European states and from much of South America.[38] The clash of the totalitarian extremes in Spain foreshadowed an end to the "twenty-year truce." For many liberals and leftists on both continents, the struggle of the Republic against fascism was the touchstone of liberal idealism. Franco's victory did not bode well for the forces of democracy.

Where despair of the democracies drove many of the intelligentsia into the communist camp, Stalin's show trials and purges shattered much of the faith in communism. For some (though hardly all) of the intelligentsia, the Soviet experiment became, as in the title of the book by six prominent writers, *The God That Failed*.[39] The cumulative effect of the double-edged extremism and the threat of war is caught by Arthur Schlesinger, Jr.: "The Soviet experience, on top of the rise of fascism, reminded my generation rather forcibly that man, indeed, was imperfect. . . . The grounds of our civilization, of our certitude, are breaking up under our feet."[40]

The Great Collapse thus amounts to the twilight of all idols. As uncertainty, ruin, and death steamed by and eclipsed the iron horse of progress, history became the agent of nihilism. The demise of progress tended to expose God's final vulnerability: His immanence in history was now denied Him. On a symbolic level, Milan Kundera captures the dynamic of the Great Collapse:

> In the course of the Modern Era, Cartesian rationality has eaten away at all the values inherited from the Middle Ages. But at the very moment of the total victory of reason, pure irrationality (the will to power) suddenly dominates the world scene, because there is no longer any generally accepted value system that could block its path.[41]

On a symbolic level, this "Second Fall of Man" instigated a hasty search for new idols.[42] I suggest that the mantle of godhood is passed on to the idea of omnipotence—ever-increasing human powers divested of the moral concerns that upheld the idea of progress.

On a practical, sociopolitical level, the Great Collapse resulted in

---

38. Hobsbawm (1987, 333).
39. Crossman (1950).
40. Cited in Bloom (1986, 188).
41. Kundera (1984, 16).
42. The term is used by Russell (1981, 72).

the virtual paralysis of the democracies, which turned inward rather than facing the political and economic threats collectively. I will examine democratic paralysis in Chapter 3, as it becomes an issue at the end of World War II and thereby helps to sanction faith in the atomic bomb.

2

The Drama of Omnipotence

The story of the atom bomb, as it is usually told, starts with revolutionary developments in physics.[1] Significant among these were the
discovery of X rays, the isolation of radium and thorium, Einstein's
theory of general relativity (which received initial confirmation in
1919), and the work of Rutherford, Planck, and Bohr on the energy
and structure of the atom. The essence of these findings, at least for
considerations involved here, is that a tremendous amount of energy
is locked up in the atom. The investigation of the atom raised few
qualms for the pioneers in this field, since it was generally thought
that its energy could not be liberated. It was as if, in constructing the
universe, God had ruled out access to its fundamental power.

The next part of the story is plotted by astonishing twists. As the

1. On the bomb, see Clark (1980); Groueff (1967); Jungk (1958); L. Lamont (1965);
Rhodes (1986); and Wyden (1985).

light changed to green at a London intersection in 1933, Leo Szilard, a Hungarian-born physicist, had a brilliant insight:

> It suddenly occurred to me that if we could find an element which is split by neutrons and which would emit *two* neutrons when it absorbed *one* neutron, such an element if assembled in sufficiently large mass, could sustain a nuclear chain reaction. I didn't see at the moment just how one would go about finding such an element, or what experiments would be needed, but the idea never left me.[2]

Indeed, at the time scientists believed that the atom was indivisible. Yet in late 1938 the uranium atom was successfully split. As Niels Bohr, the Danish physicist and Nobel laureate, observed, this was "at variance with all previous experience in nuclear physics." Moreover, it soon became clear that a chain reaction was feasible. The scientists had peeked at God's secret. Still, interminable difficulties stood in their way, and there was no certainty that the secret could be pried open.

Then comes the motivation. According to the historian Ronald Clark, the "drama would have unfolded in a different manner had it not been for an event which had taken place in 1933: the appointment of Adolf Hitler as Chancellor of Germany on 31 January." Clark even suggests that the "search for the greatest power on earth might have been, if not abandoned, then at least put into cold storage until those involved had thought a little longer about the Pandora's box they might be opening."[3] So Hitler provided the impetus for an event that otherwise might not have happened. This is an admirable hope, but one that I argue is without foundation.

The next installment takes us into the labyrinth of the Manhattan Project, the secret and massive American effort to build the bomb that culminated in the successful Trinity explosion. That success is followed by the decision to employ it against Japan; then the horrors of Hiroshima and Nagasaki are described. Often the story ends here. In other cases, the Soviet atom bomb, the hydrogen bomb, and the arms race are traced out in further installments.

This is our story, yet it is not our story. Chronicles built on this structure are all too likely to take the bomb for granted. It is explained primarily by its immediate antecedents: the confluence of startling

2. Cited in Clark (1980, 35).
3. Clark (1980, 41, 62).

scientific breakthroughs with a total war against a malefic foe. The atomic age is another rung in the ladder of scientific progress, albeit a step so large that it is simultaneously awe inspiring and daunting, an unintended fusion of fate and terror.

What is missing from this is a radical and thorough sense of the bomb's uniqueness. Einstein's admonition—"the atom has changed everything except our modes of thinking"—is pertinent not only to the aftermath of Trinity, a lineage that now spans more than four decades, but also to our understanding of the historical factors that placed the bomb on the human agenda. A radical reappraisal of the possibility of the "gadget" (it was not referred to as a bomb) raises many questions. What made the gadget *thinkable* in the first place? What gave rise to the *mentality* that allowed humanity to rummage in God's secret chambers? What, in the final analysis, rendered the bomb so *inevitable?* At issue here is not just a chronology of scientific advances, but a sense of hubris, of total dominance and mastery of the physical world that emerged as God's powers were replaced by the faith in reason and progress. At issue is not just a war with Germany, but a host of social, economic, and political conditions that facilitated and actively promoted, both speculatively and practically, humanity's tinkering in the unknown.

Once the fateful step was taken, other issues came to the fore. What imparted such an impetus to the sprouting of nuclear arsenals? And why did this growth continue even as experience demonstrated that the arsenals are as prostrate as they are lethal?

There are, then, many unasked questions. Robert Lifton underscores unwritten volumes when he declares that "the bomb impairs our capacity to confront the bomb."[4] Indeed, where are our nuclear icons? So absurd and terrifying a reality should provoke our symbolic temperament! Whereas an outpouring of icons might well be expected, perhaps the only good candidate for nuclear imprinting is the mushroom cloud (itself a universal and potent symbol).

Lifton has pursued the need to renovate our thought and imagination. To discern the effects of the nuclear reality, he suggests that we "require a model or 'paradigm'—an overall 'controlling image' " that would be equivalent to instinct in the Freudian paradigm.[5] To give mythical substance to our shadowy images, he unfolds an array of symbolizations of death. Starting with "images of extinction," he

4. Lifton and Falk (1982, 3).
5. Lifton and Falk (1982, 62).

painstakingly draws out the ways in which the "new ephemeralism" and the possibility of a "break in the human chain" threaten our sense of immortality—"part of the universal inner quest for continuous symbolic relationship to what has gone before and what will continue after our finite individual lives."[6]

The Fate of the Earth, by Jonathan Schell, is also a concerted effort to renovate thinking about our nuclear predicament. Like Lifton, he focuses on images of extinction. Our nuclear progeny will be termites and crabgrass: "Death, having been augmented by human strength, has lost its appointed place in the natural order and become a counter-evolutionary force."[7] The "second death" will cancel the future for all unborn generations. From our common menace, he then deduces principles of life for the new common world.

Both Schell's work and the postulation of "nuclear winter" revived nuclear fear in the early 1980s and probably made a substantial contribution to the growth of the peace movement in that period. But if the fear of extinction grasped the public consciousness then, it had been preceded by a near-panic over Soviet nuclear developments—the window of vulnerability, which was used to explain their adventurism in Africa and Afghanistan.[8] And before this, the public culture surrounding the bomb can best be characterized as one of passivity; from the end of the Cuban missile crisis until the Carter presidency, "nuclear forgetting" reached new heights.[9] Nuclear issues of course persisted; but they no longer commanded the degree of attention and *emotional energies* characteristic of other periods.

Along with nuclear fear and nuclear forgetting, there is also *nuclearism,* the faith in the bomb as a winning weapon and a symbol of America's uniqueness. Images of extinction, when they are not subject to numbing, do characterize the nuclear world; but so do images of life. The bomb has long been regarded as the preserver of Western society, as the "only thing it had" to ward off the communist threat. Thus few things (if any) have been as sacred as the American deterrent. In the same vein, the announcement of Star Wars can be conceived as an effort to reverse the growing fear of extinction by promising to regenerate American omnipotence. Star Wars aimed to conquer the *social* death, to provide a canopy that would assure the perpetuity of the social order and thereby afford a symbolic sense of historical transcendence to the individual.

6. Lifton and Falk (1982, 64).
7. Schell (1982, 113).
8. I will discuss this in more detail in Chapter 9.
9. See Boyer (1980).

The Feeling of the Numinous

Given these contrasts and the radical shifts they have embodied, I suggest here that the bomb can best be grasped as a numinous phenomenon. I begin by discussing Rudolf Otto's conception of the numinous insofar as it is relevant to the bomb.[10] In particular, I stress the ambivalent and thereby confounding characteristic of the numinous; in the case of the bomb, it gives rise to the pressing need to control and assuage the fear associated with this power. Ambivalence is not just the property of individual minds, but is often present at aggregate levels, existing as a part of the public culture.[11] My analysis suggests that the social patterning and determination of ambivalence is central to an understanding of nuclear weapons.

According to Otto, the numinous is the feeling of the holy that remains when the concept fails. Said another way, neither the deity nor the religious state of mind is exhausted by rational attributes; they also involve an "overplus of meaning." The experience of the holy, which constitutes the "real innermost core" of religion, entails feelings and experiences that are ineffable and irreducible. The holy must be awakened and felt, not taught. By implication, efforts to apprehend the numinous through concepts can only amount to a schematization of a "unique original feeling-response."

Grasped conceptually, the numen combines a twofold consciousness, or feeling response: on the one hand, feelings of "absolute unapproachability"; on the other, feelings of "absolute overpoweringness." Absolute unapproachability gives rise to *creature feelings*, constituted by the "submergence into nothingness before an overpowering, absolute might of some kind."[12] Allied with this is a feeling of *mysterium tremendum,* denoted by such indicators as "daemonic dread" and "awe," which unfold as "something uncanny," "eerie," or "weird." The numen is "wholly other," that which is "beyond the sphere of the usual, the intelligible or the familiar."[13] It evokes a sense of "stupor," striking us dumb with absolute amazement.

But the other pole of the numen, characterized by the feeling of absolute overpoweringness, gives rise to a sense of *fascination.* Further:

> These two qualities, the daunting and the fascinating, now combine in a strange harmony of contrasts, and the resultant

10. See Otto (1958) and Chernus (1986, 12–31).
11. On ambivalence as an aggregate phenomenon, see Room (1976).
12. Otto (1958, 10).
13. Otto (1958, 26).

> dual character of the numinous consciousness, to which the entire religious development bears witness, at any rate from the level of the "daemonic dread" onwards, is at once the strangest and most noteworthy phenomenon in the whole history of religion. The daemonic-divine object may appear to the mind an object of horror and dread, but at the same time it is no less something that allures with a potent charm, and the creature, who trembles before it, utterly cowed and cast down, has always at the same time the impulse to turn to it, nay even to make it somehow his own.[14]

Effectively, there is an ambivalence composed of a sense of daemonic dread or unnatural fear that is simultaneously accompanied by attraction to an omnipotent power that one seeks to master. Or as Chernus views it, "The Bomb, in short, is a symbol not only of annihilation but also of power without limit—of omnipotence—and hence of numinous mystery."[15]

Otto approaches the numinous from a developmental perspective. The lower stage of numinous consciousness is characterized by daemonic dread. With the Hebrew Bible and then the New Testament, the numinous was rationalized and moralized, but not superseded.[16] Even in the Middle Ages and beyond, the Christian God, whose attributes included omnipotence, goodness, wisdom, and truth, still could display the "awefulness," "fury," and "wrath" of His nonrational divine nature. Both the Black Death of the fourteenth century and the Lisbon earthquake of 1755 often were felt as terrifying manifestations of His numinous wrath.

But the dominant tendency in Christianity was rationalization, which ultimately excluded most nonrational, and hence numinous, elements from religion. What remained of the idea of God was "a one-sidedly intellectualistic and rationalistic interpretation."[17] This reduction of the deity to rational attributes parallels the process of disenchantment discussed in Chapter 1.

The reviving of the feeling of the numinous by the bomb is both paradoxical and intriguing. Modern science was one of the main carriers of disenchantment, "freeing" the natural and the social worlds from "metaphysical" and "religious" assumptions. Yet it was through the agency of science that a sense of the numinous was

14. Otto (1958, 31).
15. Chernus (1986, 15).
16. Otto (1958, 111).
17. Otto (1958, 3). On the rationalization of Christianity, see also Berger (1969).

revitalized. As a *human* creation, the bomb was a unique source of overwhelming and limitless power. In prying open the secrets of the atom, humans were in effect recouping some of the powers that they ascribed to the departed deity. The scientists who created it shared, as Creators, its numinous aspects; the power of the bomb reflected and embodied human powers. Yet the bomb and its associated human powers still evoked creature feelings. According to Robert Oppenheimer, who was scientific leader at Los Alamos and sometimes named the "father" of the bomb, the atomic scientists, in creating it, had "known sin."[18] Like those who had eaten the forbidden fruit of the Tree of Knowledge, mortals had released forces too great for them to manage. As this daunting power fell into human hands, the fear of God was replaced by the fear of humanity, and we may never have feared God as much as we fear ourselves.[19]

Entering the Workshops of the Powerful

The drama of unleashing the first nuclear explosion in history was unsurpassed. There was the malevolent enemy, Nazi Germany, with a lead of up to a year and a half in the building of a bomb. There was the even greater spectacle pitting the knowledge and ingenuity of scientists against a clandestine nature. There was the largest project ever undertaken, done in utmost secrecy during the course of an

18. Cited in Weart (1988, 113). See Kaufman (1985) on the idea of the bomb being "out of sync" with human powers.

19. In applying the idea of numen to the bomb, there is the inevitable question of the "fit" between the two. To put it another way, how much is real and how much is metaphor? Here the sociology of religion is of little help. Transcendental religious phenomena have been so problematic for the scientific study of religion that they typically are ignored (as in Berger's [1969] "methodological atheism") or interpreted functionally and thereby divested of any supernatural associations (Garrett, 1974). Here I suggest that we are not overextending the concept, insofar as the bomb did encompass the two poles of the numen—daunting fear and fascinating power. Now the power of the bomb proves not to be omnipotent, but this does not gainsay the fact that it was *perceived or felt* to be omnipotent. In the primitive religions that Otto discusses, it was common to attribute extraordinary powers to plants and animals that presumably held no such powers; it was the perception that mattered. The greatest change in the concept is that the power (or perception of omnipotence) now fell into human hands. This change renders the bomb extraordinarily problematic, but it does not invalidate its numinous standing. At the same time, we lack a language commensurate with the nuclear experience, and over the years we have inevitably routinized the bomb. See also Chernus (1986, 17, 26–27).

exacting war. And then there was the danger, the specter of a life-exterminating holocaust. Here, in other words, was the ultimate test of human omnipotence. Success would be an unabashed testimonial to the possession of transcendent power.

Right from the start of this enterprise most scientists realized that they were not simply creating a much bigger bang. They were toying with one of the ultimate forces of nature. Yet in the end, these scientists were as much the victims as the initiators of the drama. They were ambivalent about the idea of building a bomb; they were stalked by moments of foreboding in creating it; the end product horrified them. For all these portents, they proceeded with the utmost haste and effort. Splitting the atom was charged with presentiments of the diabolical. Splitting the atom also allowed scientists to rise to the empyrean. The uranium pile was the philosophers' stone. They were caught up in the drama of the Manhattan Project and even more so in the testing and display of their own omnipotence.

This is the first part of our story: the moral drama of omnipotence that gave such impetus to and rendered indispensable the bomb. Where histories of the bomb usually focus on patriotism and the threat of the Nazi atomic program, I suggest that these motivations were (or became) secondary; they were certainly no greater than the scientists' desire to test their own powers. Thus our story focuses on the godlike posture the scientists were assuming. This includes their tampering with taboo forces that might extinguish life; their Genesis Complex—their desire to create a new world regardless of the consequences; and the unprecedented resources put at their disposal so that, as apprentice deities, they could assay their claim on godhood. Where Laplace's Divine Calculator never managed to predict the future, the physicists re-created it.

After the war, Werner Heisenberg, a leading German scientist, stated, "In the summer of 1939 twelve people might still have been able, by coming to mutual agreement, to prevent the construction of atom bombs."[20] Heisenberg actually approached Bohr in September 1941 with such a suggestion. But the world of physics, perhaps the last remnant of the internationalism associated with the idea of progress, was now rife with suspicion, and Bohr thought he was being told of advances toward a Nazi bomb.[21]

The race for the bomb started even before the outbreak of war. In

20. Cited in Clark (1980, 62–63).
21. See Wyden (1985, 84–86).

Germany, nuclear research was consolidated under the War Office, uranium exports banned, and an effort made to corner the available heavy water. In France, Frédéric Joliot-Curie initiated a program to build a bomb in the Sahara. Britain began its project in 1939 and soon had a substantial lead. The soviets accelerated their nuclear program in 1939 and moved it into the interior following the German invasion. In the United States, a Uranium Committee was formed in October 1939.

Contrary to Heisenberg and his twelve people, it was the scientists who were the entrepreneurs behind the crusade for atomic power. They had to overcome bureaucratic and military conservatism, the belief that no new weapon could be decisive, and the focus on immediate prospects, like radar, rather than the uncertain promise of some magical device. In Germany that resistance was never overcome; consequently, and in conjunction with an error that led scientists to select the wrong moderator, the bomb project was stillborn. In the United States, the scientists' crusade was initially impelled by fear. Its leaders were mostly Jewish refugees who worried that a German nuclear lead would culminate in their unprincipled use of atomic bombs. Despite the dire warnings, it took three more years before the essentially moribund Uranium Committee was transformed into the Manhattan Project, in September 1942. While the sense of panic makes good drama in bomb histories, it was not decisive in the United States. Indeed, the country was at war almost a year before the project was initiated (the United States was hardly threatened by a bomb since no German aircraft could reach it; thus, in his famous letter to Roosevelt, Einstein warned of a ship carrying such a weapon). In making the decision, Roosevelt looked beyond the German threat (to nuclear power) and seemed "surprisingly indifferent to assessing it."[22]

The race for the bomb thus represented far more than a deterrent to a Nazi atomic program. In 1933 Rutherford said that researchers investigating the nucleus of the atom were not "searching for a new source of power, or the production of rare and costly elements. The real reason lay deeper and was bound up with the urge and the fascination of a search into one of the deepest secrets of nature."[23] In the Manhattan Project, this deep motivation was retained and much expanded. According to Oppenheimer:

Almost everyone knew [the project] was an unparalleled opportunity to bring to bear the basic knowledge and art of science

22. This paragraph draws on Rhodes (1986); see especially chapters 12 and 13.
23. Cited in Clark (1980, 34).

for the benefit of his country. Almost everyone knew that his job, if it were achieved, would be a part of history. This sense of excitement, of devotion and of patriotism in the end prevailed.[24]

Patriotism, yes, but it was bettered by the excitement of creation, spawned by what Oppenheimer termed "our rather blasphemous sense of omnipotency." In a recent autobiography, Victor Weisskopf muses:

> Today I am not quite sure whether my decision to participate in this awesome—and awful—enterprise was based solely on the fear of the Nazis beating us to it. It may have been more simply an urge to participate in the important work my friends and colleagues were doing. There was certainly a feeling of pride in being part of a unique and sensational enterprise. Also, this was a chance to show the world how powerful, important, and pragmatic the esoteric science of nuclear physics could be.[25]

Almost a half century before the bomb made it concrete, there was a nascent sense of scientific omnipotence identified primarily with the ideas of Einstein. The advance of the machine-based civilization had been carried forth by "ingenious tinkerers," drawn to practical rather than theoretical knowledge. For the most part, progress moved at a level people could grasp. The rudiments of the mechanistic outlook were widely shared, and science and common sense overlapped at significant points. There was no need to understand Newton or the laws of the conservation of energy to come to terms with the steam engine. But the new physics changed this. In 1906, one year after Einstein published his theory of relativity, the *Nation* asserted:

> Today, science has withdrawn into realms that are hardly [intelligible]. . . . Physics has outgrown the old formulas of gravity, magnetism and pressure; has discarded the molecule and atom for the ion, and may in its recent generalizations be followed by an expert in the higher, not to say the transcendental mathematics. . . . In short, one may say not that the average cultivated man has given up science, but that science has deserted him.[26]

24. Cited in Clark (1980, 185).
25. Cited in Bernstein (1991, 48).
26. Cited in Kevles (1978, 98).

The discovery of X rays and radium created a vision of a technological Garden of Eden. They were applied to everything: bones, welds, cancer. Crowds lined up "to see their bones," and possession of a personal X ray became a status symbol. Radium was thought to improve health, and it was applied to clock dials to make them luminous. But the dark side of the atom was manifested quickly, the result of the sickness that befell the so-called radium-dial women. Still, nothing captured the public imagination so much as the promise of nuclear energy. Atom-smashing machines with magnets weighing up to eighty-five tons fascinated a people already enamored of colossal devices. Neutron therapy for tumors was the latest addition of physics to the medical arsenal. The media were full of mythic imagery of the coming "Golden Age." Energy, not ethics, stood at the center of the new civilization. There were promises of energy so cheap "it isn't worth making a charge for it" and atom-powered airplanes for everyone. In 1936, the *Nation* observed: "More power to the magicians. The politicians evidently cannot save us. Perhaps the scientists can."[27] Indeed, it was the growing sense of scientific omnipotence that caused scientists' claims about a magical weapon to be believed and such extraordinary resources to be put at their disposal.

Los Alamos cast scientists in a new role. Newton and his predecessors were certainly involved in utilitarian matters.[28] Ever since, science has been linked to practical affairs, military and otherwise. Historically, much of the support for science and many of its significant discoveries were primed by warfare. American and European scientists made important contributions to World War I, although none of them was decisive, as the military was generally slow to pick up on the innovations. The art of war was in many respects traditional and conservative. Inventions made in one war were usually not fully exploited until the next.

Centuries earlier, alchemists had displayed concern about the effects of their efforts. One of their mottos was: "Deny the powerful and their warriors entry to your workshops. For such people misuse the holy mysteries in the service of power."[29] This idea was not foreign

27. Cited in Kevles (1978, 269).
28. The idea of control is not inherent in science. Based on a study of Leonardo's papers, Rossi (1970, 28) concludes that "he was interested in machines more as the result and proof of human intelligence and genius than as a means of the actual mastery of nature." Machines were used more for entertainment and diversion than for production. See also Leiss (1972).
29. Cited in Jungk (1958, 3).

to modern times. At the end of World War I, George Ellery Hale, foreign secretary of the National Academy of Sciences, sought to ban the Central Powers' scientists from postwar professional societies. War guilt was attached to them for their work on poison gas and submarines. Not until 1926 did the International Research Council lift the ban on scientists of the former Central Powers. The ban would undoubtedly have been more extreme if not for the open international policy of the Wilson administration.[30]

As the nuclear physicists realized the alchemists' dream in the uranium pile, they did not guard their secrets and keep their power out of the hands of those not trained to use it. They offered their secrets up. Indeed, the scientists were part of what is routinely described as the most extraordinary scientific enterprise in history. The Manhattan Project was unique in all respects: size, cost, priority, effort, the secrecy, the site, the number of scientific impresarios, the role of scientists in an enterprise run by the military. So unique was the program that it has in many respects become sacred. Articles and books about the project are unanimously reverential; the body of mythology that has grown up around it serves to consecrate American power and the American way of life. The words of William Laurence, the official journalist of the project, are typical:

> To behold these atomic power plants standing in their primeval majesty is one of the most terrifying and awe-inspiring spectacles on earth today. There is not a sign, not the slightest hint, that within these huge man-made blocks titanic cosmic fires are raging such as have never raged on earth in its present form. One stands before them as though beholding the realization of a vision such as Michelangelo might have had of a world yet to be, as indescribable as the Grand Canyon of Arizona, Beethoven's Ninth Symphony or the presence "whose dwelling is the light of setting suns."[31]

In the end, more than two billion dollars was spent so that the scientists could test the validity of their calculations. As Herbert York observed, if the relative size of the economy is taken into account, more was spent on the Manhattan Project than on Project Apollo, which sent men to the moon.

30. See Kevles (1978, 139–54).
31. Laurence (1946, 163).

Playing Dice with the Universe

For scientists, a potent motivation was the drive to unleash—even at the risk of not controlling—nature's ultimate power. Long before the Trinity explosion, scientists mused over the possibility that a nuclear blast might generate enough heat to burn up the atmosphere. As early as 1903, Ernest Rutherford playfully suggested that "could a proper detonator be found, it was just conceivable that a wave of atomic disintegration might be started through matter, which would indeed make this world vanish in smoke."[32] Two decades later the Nobel Prize winner Francis Ashton warned against "tinkering with angry atoms," and noted that if the hydrogen in the atmosphere were transformed into helium "this most successful experiment might be published to the rest of the universe in the form of a new star."[33]

Fear of a nuclear disaster was a fateful presence as scientists precariously exposed the atom's volcanic mysteries. To take one example, preliminary estimates by Edward Teller revealed that the explosion of a bomb would liberate enough heat to ignite the atmosphere. According to Oppenheimer, "They were looking at a mathematical model for the end of the world."[34] Revised calculations revealed a probability of just less than the three in one million that the physicist Karl Compton had set as the critical value to determine whether to continue the project. His logic: "Better to be a slave under the Nazi heel than to draw down the final curtain on humanity."[35] Documents declassified in 1975 showed that scientists had estimated that a safety factor of 60 prevented ignition of the atmosphere.

But the magnitude of the event transcended the comfort of statistical probabilities. As the 16 July 1945 Trinity test drew nearer, an uneasiness was present. Lansing Lamont describes it as follows:

> The GI's and scientists at Trinity had long discussed among themselves what might happen if all calculations proved wrong and the bomb turned out to be infinitely more powerful than anyone had imagined. . . . They hid their fears in jokes and banter about how the explosion might tilt the earth off its axis or affect its solar course. They joshed about the possibility of a runaway explosion igniting the atmosphere and encircling the

32. Cited in Szasz (1984, 56).
33. Cited in Szasz (1984, 56).
34. See Szasz (1984, 50).
35. Cited in Szasz (1984, 51).

> globe in a sea of fire. . . . Edward Teller . . . spent hours
> conjecturing science-fiction fantasies and situations that an
> atom bomb might produce if undiscovered laws of nature were
> suddenly activated by its explosion. Always there was the
> haunting thought that the test blast might introduce some
> violent phenomenon beyond the ken of the scientists.[36]

On the night before the test, Fermi took wagers on how long it would
take for the earth to be incinerated in the event that the atmosphere
was ignited. As the ball of fire rose in the desert, a military officer
exclaimed, "The longhairs let it get away from them."[37]

Hindsight reveals that the act was not reckless. It was a vain act,
however. Unlike the original Deity, His substitute does play dice with
the universe. The dice were loaded by theory and calculations. But
the loadings were not exact. The scientists greatly underestimated the
size of the first blast. General Leslie Groves, the military director of
the program, wrote in May 1945 that "the responsible heads at Los
Alamos felt that the explosive force of the first implosion-type bombs
would fall somewhere between 700 and 1500 tons [of TNT]."[38]
Oppenheimer guessed that it would be the equivalent of 300 tons,
and most other scientists made similarly low estimates. Such magni-
tudes were strikingly larger than the 10-ton blockbuster then in use.
Yet estimates made at the Trinity explosion revealed an actual yield
close to 20,000 tons! Scientists also underestimated the number of
deaths that would be caused by the first bombing: the figure of 10,000
pales as compared to the actual deaths at Hiroshima of approximately
70,000 (Japanese estimates are on the magnitude of 140,000). Further, they failed at first to comprehend the long-term effects of the
fallout created in atomic explosions, as they were to learn in the
aftermath of Hiroshima. Initial Japanese reports of radiation poisoning were vehemently denied by Groves, who termed them a "hoax or
propaganda" contrived to win international sympathy. American
scientists believed that, to suffer radiation damage, people would have
to be so close to the blast that they would be killed by its force alone.
For all this, it is conceivable that they could have underestimated or
missed phenomena with even more horrendous consequences.

The stories of tampering with taboo forces—and their recounting

36. L. Lamont (1965, 129–30).
37. Cited in Szasz (1984, 60).
38. Cited in Clark (1980, 204).

in awed tones in bomb histories[39]—are less about recklessness than dramatic theatrics, highlighting the command over the elemental forces of nature by scientists who were mimicking the deity. For the first time, humans were creating the "theater" of the apocalypse that would eventually result in "exterminism."[40] The scientists themselves—and later their public—were awestruck by their ability to trifle with and command such forces. The ultimate issue was put succinctly by Hans Bethe: "But will nature act in conformity with our calculations?"[41]

That the intellectual contest and promise of power overrode fear of a Nazi bomb as motivators is indicated by what Richard Rhodes terms one of the "mysteries" of the war: the American failure to mount an early and dedicated espionage effort to chart German progress.[42] Such an effort was not undertaken until late 1943. When it was known for certain by the end of 1944 that Germany had no atomic bomb, Samuel Goudsmit, scientific head of espionage, wrote, "Now that the States are in no danger from a German A-bomb, they'll leave their own in cold storage."[43]

Goudsmit's contention had merit, both technically and morally. The bomb was initially built as a form of security against an evil power with no inhibitions against using it. That original motivator was now gone. The Manhattan Project could be terminated just short of success and, with a combination of luck, effort, and desire, the gadget mothballed. But the scientists did not even begin to pursue this path. They were at the brink of omnipotence, arming the catapult. As Oppenheimer observed, they had to test "because there were too many uncertainties." According to Philip Morrison, "Up to Trinity the whole thing was an intellectual game. . . . For us, the test was the climax of everything."[44]

Politically, militarily, and scientifically, Goudsmit's musings were a fantasy. Groves had been charged with an enterprise that could "win the war." If ready—and everything possible was done to make certain it was—the bomb would be used. Well before Trinity, the gadget was

39. My reading suggests that bomb histories have become less reverential over time, but have not abandoned that attitude. The mythology still takes precedence over its debunking. But then the Manhattan Project was impressive. As Bohr would say, "You see, I told you it couldn't be done without turning the whole country into a factory. You have done just that" (cited in Rhodes [1986, 500]).

40. The term is used by Thompson (1980).

41. Cited in L. Lamont (1965, 144).

42. Rhodes (1986, 605).

43. Cited in Bar-Zohar (1967, 50).

44. Cited in L. Lamont (1965, 298).

already being cast as indispensable—to justify the expense and effort of the Manhattan Project, to defeat Japan without an invasion and thereby save American lives, as a scientific experiment on the effects of a nuclear blast, as an indirect warning to the Soviets.

Only Leo Szilard would come close to Goudsmit's position. "I began to ask myself," Szilard would say, "What is the purpose of continuing the development of the bomb, and how would the bomb be used if the war with Japan had not ended by the time we have the first bombs? . . . It was not clear what we were working for."[45] As I will show in Chapter 3, Szilard did protest against use of the bomb on Japan and the American effort to maintain "secrecy"; neither he, nor Bohr, who mounted his own campaign against secrecy, sought to halt the building and testing of the bomb itself.

Each group had some of its own reasons for desiring the rapid completion of the bomb. But they all coalesced around the continuity of the project and the importance of the gadget. Goudsmit was ultimately seeking to *stop* a German bomb. He had no sense of the drama involved in *creating* an American bomb. Like Prometheus, the scientists were seeking to steal a fire hitherto known only to the gods.

Trinity, the name given to the site constructed for the initial test, has created as much controversy as practically anything in the project. Oppenheimer was to say that "this code name didn't mean anything." He had been reading a sonnet by John Donne, and that name had popped into his head:

> Batter my heart, three-person'd God; for, you
> As yet but knock, breathe, shine, and seek to mend . . .

But what significance has been read into the name! Lamont is typical:

> Was it blasphemy to christen with such a name the birthplace of the atomic bomb? Perhaps. And yet there was a hint of holiness about the silent valley that was Trinity—something in its history that touched the indomitability of man's questing spirit; something in its ageless and primitive beauty that suggested man's earliest temples of prayer.[46]

Others, in contrast, have suggested that Trinity has Hindu connotations. Oppenheimer taught himself Sanskrit and was familiar with the

45. Cited in Kurzman (1986, 262).
46. L. Lamont (1965, 73).

Hindu concept of trinity, consisting of Brahma, the creator, Vishnu, the preserver, and Shiva, the destroyer. In the Hindu cycle, things are transformed, not destroyed.

In *The World Set Free,* a novel published in North America in 1914, H. G. Wells predicted the advent of atomic energy. The day after the atomic breakthrough, Holsten, Oppenheimer's fictional counterpart, felt that he was "something strange and inhuman, a loose wanderer from the flock returning with evil gifts from his sustained unnatural excursions amidst the darknesses and phosphorescences beneath the fair surfaces of life."[47] The reactions to the Trinity test of Monday, 16 July 1945, were along similar lines. According to Victor Weisskopf, "Our first feeling was one of elation, then we realized we were tired, and then we were worried."[48] Scientists hugged, formed a spontaneous chorus line and did a snake dance, and one after another screamed joyously over the PA system.

It was only in the aftermath of the Trinity explosion that the inchoate sense of omnipotence was fully experienced as the feeling of the numinous. Just before the explosion concern was directed at dangers like tilting the earth off its axis. After it occurred, scientists and other participants drew on the arcane language of alchemy or the mythology of the Bible or other religious texts to describe its awesome power. Descriptions combined a reverence hovering on the religious with a sense of foreboding bordering on the diabolical. The scientists, in essence, were seeking a language commensurate with the feeling of the numinous.

Of the many documented statements, I cite a few. General Thomas Farrell's account was more than military:

> The effects could be called unprecedented, magnificent, beautiful, stupendous, and terrifying. No man-made phenomenon of such tremendous power had ever occurred before. The lighting effects beggared description. . . . Thirty seconds after the explosion, came, first, the air blast pressing hard against people and things, to be followed almost immediately by the strong, sustained, awesome roar which warned of doomsday and made us feel that we puny things were blasphemous to dare tamper with the forces hithertofore reserved to the Almighty.[49]

47. Wells (1914, 36).
48. Cited in Szasz (1984, 91).
49. Cited in Groueff (1967, 355).

George Kistiakowsky, a scientist, was more succinct: "I am sure that at the end of the world, in the last millisecond of the earth's existence, Man will see what we have just seen."[50] Test Director Kenneth Brainbridge called it a "foul and awesome display" and told Oppenheimer, "now we're all sons of bitches."[51] Oppenheimer quoted the Bhagavad Gita: "I am become death, the shatterer of worlds." Stan Ulam, who chose not to witness the blast, remembered those who returned: "You could tell at once they had a strange experience. You could see it on their faces. I saw that something very grave and strong had happened to their outlook on the future."[52]

Humans had once before tried to scale the heavens. But at the Tower of Babel, God had thwarted their efforts by causing them to speak different languages. But He apparently overlooked the "universal language" of science.[53] In the half century of scientific work leading up to Trinity, many demons had been encountered and overcome. Reality had sided with the calculations. The desire to test themselves had helped the scientists exorcise each demon. Their final performance had a godlike magnitude. They were Creators, and in more than one way. But as they felt the shock waves that initiated a new world, their ambivalence became palpable. Only after Trinity could they truly see the forces they had unleashed. With their brilliance they had become death's collaborator. They had a hold on transcendental power—a power as diabolical as it was divine. But the final demon could not be exorcised, for it was all-too-quickly usurped by others.

50. Cited in L. Lamont (1965, 244).
51. Cited in Szasz (1984, 90).
52. Cited in Szasz (1984, 91).
53. Schell (1982, 103).

3

The Testing and Display of Indispensable Power

With the Trinity explosion, the bomb passed into the hands of the politicians. They already had, so to speak, their fingers reaching out for it. As Oppenheimer observed:

> After the collapse of Germany, we understood that it was important to get this ready for the war in Japan. We were told that it would be very important to know the state of affairs before the meeting at Potsdam at which the future conduct of the war in the Far East would be discussed.[1]

For the politicians, as with the scientists, the sense of overwhelming power came first and early. Nuclear dread, the disquieting wrath and

1. Cited in Feis (1966, 41).

taboo associated with that power, came in its wake. This, in turn, was followed by efforts to overcome the reaction of fear, primarily through an attempt to construe the bomb as a moral bequest that was proof of America's special standing among the nations of the world. For the politicians, as with almost everyone else, the bomb would actually prove to be a "confounding" power, and thereby generate competing symbolic and practical efforts to manage it.[2]

The early sense of power is clear in Lord Cherwell's 1941 warning to Churchill that bomb work might be too valuable to transfer to the Americans:

> The reasons in favor [of an English location] are . . . above all the fact that whoever possesses such a plant should be able to dictate terms to the rest of the world. However much I may trust my neighbor and depend on him, I am very much adverse to putting myself completely at his mercy.[3]

Cherwell was certainly voicing the attitude that Churchill would adopt to the bomb, and it was characteristic as well of many (but not all) American politicians.[4]

That politicians were prepared and indeed eager to attribute such power to this unproved device had, as I suggested earlier, much to do with the Great Collapse. Chapter 1 provided a broad overview of the Great Collapse that focused on its symbolic aspects, as well as on the economic, military, and political enemies at the gates. Here the emphasis is on one consequence of these threats: democratic paralysis. The absence of an effective balance of power and collective economic solutions resulted in drift and paralysis. Thus at the start of World War II Roosevelt was so pessimistic about the future that he felt that all the United States could do was to be a "citadel" in which Western civilization "may be kept alive."[5] Most important, the analysis suggests that the sociopolitical factors that undermined balancing in the interwar years either carried through the war or appeared likely to reemerge as the conflict ended. It is in this context of persistent threat and the feared absence of an effective postwar balancing system that Western political leaders invested their faith in the bomb.

2. See Beckford (1983).
3. Cited in Rhodes (1986, 372).
4. See Sherwin (1977, 83).
5. Roosevelt (1941, 150).

The Failure of Collective Action

George Kennan wrote, "There is, let me assure you, nothing in nature more egocentrical than the embattled democracy."[6] During the 1930s the democracies met political and economic crises by turning inward and foregoing collective action. Only when events portended a threat to their very survival did national and other sectional interests give way to a real concern for collective security. The failure of the democracies to act concertedly is clear.[7] Here I wish to indicate the extent to which this failing was not just a result of inept leadership or the misreading of obvious threats, but had structural roots that also rendered unlikely collective action and security after World War II. In other words, the overlapping and intersecting threats that constituted the Great Collapse gave rise to a democratic paralysis that seemed likely to reemerge after the war.

Collective action is used here to encompass both the conventional balance of power and multilateral agreements that, unlike balancing, are broader and not aimed simply at thwarting bids for hegemony.[8] Thus the depression necessitated multilateral agreements to stabilize currencies and facilitate trade. Most states, however, concentrated on domestic economic management and developed a rival attitude toward other states that bordered on autarky.[9] Domestic economic concerns, coupled with the effects of the depression on military preparedness, in turn militated against the operation of balancing.

Inaction is the rule for international threats. As far as possible, states avoid confrontations since they recognize the risk of events escalating out of control.[10] Inaction was also induced by competing conceptions of national interest. Thus the British-French alliance lapsed in the early 1920s, as the former was concerned with empire and the latter with Germany.[11] American isolationism, coupled with general antipathy to the Soviets, effectively removed regimes that might have bolstered alliances and the League of Nations. Only concatenating disasters led the new Roosevelt administration to recognize the Soviets in 1933.

In the Europe of the 1930s, a number of additional factors induced

6. Kennan (1960, 5).
7. On the failure of collective action and/or appeasement, see Divine (1962); Jonas (1966); Kennedy (1988a); Offner (1969); Rock (1977); Ross (1983); and Taylor (1964).
8. See Waltz (1979).
9. Kindleberger (1973).
10. Schelling (1966).
11. Marks (1976).

paralysis. Above all, the totalitarian extremes that rent the continent were frequently reproduced in the class conflicts within each state. This class polarization, which inhibited concerted action, was exacerbated by the depression, and it stoked the fear, given credence by the overthrow of the czar, that behind war lurked revolution. Thus in the Spanish civil war, the French popular-front government perceived a risk of internal conflict if it defied rightists and aided the Republic.[12] Collectively, all the democracies could agree upon was to remain neutral. The "volunteer brigades" that went to Spain aptly characterize their stance.

The political elite was also typically split, some (usually the minority) calling for forceful responses and others for toleration and negotiation. As they had to contend with shifts in both the international context and popular sentiment, they sometimes changed their own positions. Thus Neville Chamberlain, British statesman and prime minister, adopted a profascist position in the mid-1930s, when communist support was at an all-time high and reconciliation with fascism seemed possible because at least it was capitalist. As the Nazi threat increased and popular opinion balked at the toleration of fascism, official support swung back to the Soviets.

Churchill, whose forecasts from 1915 on were probably more accurate than those of anyone else, complained that the democracies suffered from a lack of nerve: "The British and French cabinets at this time [the 1938 Czechoslovakian crisis] presented a front of two overripe melons crushed together; whereas what we needed was a gleam of steel."[13] Indeed, with the memory of World War I and with the domestic concerns brought on by the depression, it is hardly surprising that the populace was not psychologically ready for entangling commitments and that efforts at reassurance worked. The economic problems of the depression also translated into military weakness, and little was done to redress this until late in the 1930s.[14] Given differences in national interest, class conflicts with international implications, the fear of revolution, shifting sentiments, efforts at reassurance, and the general desire to avoid a war they were not militarily prepared for, it is hardly surprising that the democracies could not act in concert.

A real problem for the democracies was that these paralyzing factors either carried over into the war or were likely to and in many cases did reemerge as the war was ending. The powers in Europe did

12. Fleming (1961, 65).
13. Churchill (1948, 310).
14. See Shay (1977).

not settle into a two-bloc formation until two years after World War II began.[15] Divergences of national interest and the limits of alliances affected both the prosecution of the war and plans for its aftermath. While Churchill often spoke of the "special relationship" between the Americans and the British, more recent histories have debunked this notion and demonstrated that coalition warfare resulted in constant rivalries and cross-purposes.[16] In particular, where British imperial interests and strategies favored a back-door offensive in North Africa and the Mediterranean, Roosevelt preferred a frontal strategy in Europe. The upshot was compromises that delayed the Allied invasion until 1944 and guaranteed that the Soviets "liberated" Eastern and Central Europe. Where France and the Soviet Union wanted Germany dismembered, the Americans and British came to favor a more stable Germany. But where Britain and France wanted to retain their empires, Roosevelt planned to seek an end to this nineteenth-century legacy. British-American divergences became especially marked as the war was ending: where Churchill sought to contain the Soviets in Eastern Europe, Roosevelt was more tolerant of their security needs and, to avoid the appearance of "ganging up" on them, distanced himself from the British. These divisions continued after the war, when, for example, Henry Wallace spoke for those who worried that British interests were governing Truman's approach to the Soviets.[17]

The end of World War II, perhaps more than the end of World War I, was attended by a sense of collapse that rendered collective action and security problematic. There was the need to reconstruct national economies devastated by the war, restore democratic institutions, and deal with the challenge of communism as well as the virtual collapse of the European balance of power. The prewar economic history of Europe did not augur well for postwar economic growth. The European economies were near collapse.[18] Low production and the food crisis of 1946 were followed by the "dollar gap" of 1947. That an economic recovery was under way by 1948 came as a complete surprise. What was termed Europe's "economic miracle" restored confidence in a capitalist system that had hovered on disaster for so long.

There was also the need to reintroduce democratic institutions, particularly in France and Italy, which seemed to be at the brink of

15. Waltz (1979, 176). On alliance limits, see Liska (1962).
16. See Eisenhower (1986) and Harbutt (1986).
17. Walton (1976, 63–64).
18. See Laqueur (1970, 7) and Yergin (1977, 305–7).

anarchy. The Communist parties, which had gained considerable prestige from Stalin's victories and their role in the resistance, polled more votes than ever. In the France of 1946, the Communists were the strongest party and, as in Italy, held cabinet positions. That the Communist parties (excepting Belgium and Greece) did not try to take advantage of possible revolutionary situations was unexpected. Also unexpected was the decline of the Communist vote after 1948 and the rise of Christian-democratic parties.

Where the temporary eclipse of German power at the end of World War I was preceded by the temporary disappearance of Russian power, World War II utterly destroyed any balance and brought Russia to the heart of Europe. The disintegration of Germany was accompanied by the all-but-complete collapse of the military power of the European democracies. The British note of 1947 telling the Americans that they could no longer help in the Greek civil war crystallized their impotence. At Yalta, Churchill had worried about the balance in Europe after Roosevelt told him that, because of congressional and public pressures, the American occupation would be limited to two years.[19] Even in the face of Churchill's distrust, it seemed inevitable that the Soviets would step in to fill the power vacuum in Central Europe.

The war left the United States as *the* world power. But as Roosevelt had anticipated, popular opinion quickly turned against entangling commitments in Europe. Demonstrations by American troops and a bring-the-boys-home campaign resulted in a faster withdrawal than planned. After the war, the American army was rapidly demobilized, war industries dismantled, peacetime budgetary constraints imposed on the military, and universal military service rejected. Isolationism reemerged as an issue in the United States after the war and was a particular source of elite conflict.[20] The postwar isolationism was not the same as in the 1930s, since the United States was no longer impregnable and the emerging Cold War was to be cast in moral terms. Although the Cold War was not yet regarded as inevitable, divergent interests among the big-three powers surfaced before the defeat of Germany and increased as the common enemy disappeared. Thus the one country capable of counterbalancing the Soviets was apparently disengaging itself. Aside from the bomb, American power was limited to its considerable but untested economic clout. Events

19. Churchill (1953, 353).
20. On the military, see Koistinen (1980) and Rosenberg (1982, 1983). On isolationism, see Sanders (1983).

would reveal that economic leverage was not effective against the Soviets.

Dictating Terms to the World

The historian Daniel Yergin observes that "the unhappy past was much with American leaders" guiding policy in the 1940s. The "close call" of the depression, the rise of dictatorships, and the two world wars "made their quest for a different kind of world order . . . urgent."[21] These leaders aimed at a Wilsonian model of international relations. However, Roosevelt sought to avoid past mistakes by grafting a more realistic conception of power on that model. Hence the United Nations would also include an executive committee made up of a consortium of great powers that included the Soviet Union. Roosevelt in fact envisioned "four policemen" capable of acting in concert outside the U.N. framework.[22]

But if the Great Collapse rendered old solutions suspect, it also undermined the trust necessary for Wilsonian internationalism. Roosevelt's "four policemen" hung on the thread of his own personal diplomacy with Stalin, which was threatened by several factors: conflicting Soviet-American interests, which lay dormant during the war; resurgent isolationism in the United States; and the ostensibly secret American atomic bomb.[23] At the same time, Roosevelt's plan was handicapped by the need for congressional approval and by American public opinion, which was antithetical to anything suggesting big-power politics.

Against the uncertainty of Wilsonian internationalism were a number of other models for structuring the postwar world that derived their sanction from the bomb. Clearly, there were too many uncertainties about this new device to give rise to any single policy. Indeed, the bomb offered a number of different possibilities for international relations that were recognized and contested both before Hiroshima and in the period of the American nuclear monopoly. These alternatives correspond in a fundamental way to the two poles of the bomb's numinous ambivalence.

Even before the gadget was known to work, nuclear dread began to

21. Yergin (1977, 9).
22. Yergin (1977, 43–48).
23. Compare Kennan (1960, 355).

motivate scientists. Many of them believed that the bomb amounted to the most significant project in history, since it threatened the continued existence of civilization.[24] Hence they now regarded it as a factor that dictated a radical restructuring of international relations. Since they knew there was no "precious" atomic secret, they in effect denied that it was a "winning weapon" that could be used to gain any advantage over the Soviets. Only unconstrained international cooperation could undo the threat it posed.

In contrast, and drawing on the power pole of the numen, the bomb was more generally construed as the "winning weapon." Gregg Herken used the term "winning weapon" in the title of his excellent book of 1980, and for good historical reasons;[25] but the term is misleading because of its narrow military connotations. The idea of the winning weapon, as used and documented by Herken, implies a device that would be decisive in the political, diplomatic, economic, *and* military realms. In other words, the bomb, unlike machine guns, tanks, or bombers, was regarded as far more than a decisive weapon of war.[26] A device capable of "vaporizing" cities was in fact invested with a *range* of winning attributes, practical and symbolic, that far transcend associations with a mere "weapon."

Diplomatically, the bomb was deemed a "decisive" power that could either bring a quid pro quo from the Soviets or neutralize potential military threats. According to Yergin, the bomb induced a "feeling of omnipotence," even though the policy elite "did not know how" to use it.[27] Somehow, it was expected to work magic by virtue of its "unstated presence." Hence there was uncertainty and a search for solutions, but all within the broad context of an omnipotent winning weapon. One version of this, consistent with much of Roosevelt's actions, was that two of the "four policemen" (the United States and Britain) would possess the bomb.[28] In addition, there were "containment" views of

24. Sherwin (1977, 48). On the scientists, see also Gilpin (1962) and Sherwin (1985). I will cover the scientists' role in more detail later in this chapter and in Chapter 4.

25. The term was also used in the Baruch Plan ("Before a country is ready to relinquish any winning weapon, it must have more than words to reassure it" [Fleming 1961, 372]) and by Senator Brien McMahon of Connecticut in 1948 ("never before had a great nation voluntarily offered to give up a winning weapon which it alone possessed" [Fleming 1961, 378]).

26. Brodie (1959) shows that before World War II bombers created a real scare and were thought capable of rapidly bringing a nation to its knees. Yet this fear has been dwarfed by nuclear fear, and bombers were never invested with the range of diplomatic and other powers ascribed to atomic weapons.

27. Yergin (1977, 121).

28. Sherwin (1977, 89).

the bomb that envisioned its use, usually diplomatically but potentially militarily, as "a possible means of controlling the postwar course of world affairs."[29]

The bomb never lived up to its promise. Indeed, it is clear that it was invested with diplomatic, political, and economic powers that it never embodied (militarily, the bomb may well have had a deterrent effect). In the remainder of this chapter and throughout Chapter 4, I examine the idea of the winning weapon in these different realms, and show how the fearful side of the bomb emerged to take its place beside the promise of omnipotence.

Roosevelt displayed an ambiguous attitude to the bomb, but his actual *decisions* suggest that he was particularly concerned with its postwar diplomatic value for the British and Americans in relation to the Soviets.[30] The importance he attributed to it is indicated by the personal control he established over nuclear policy.[31] According to Vannevar Bush, Roosevelt's policy in 1944 seemed to be to "join with Churchill in bringing about a US-UK postwar agreement on this subject [the atomic bomb] by which it would be held closely and presumably to control the peace of the world."[32] To a large extent he was influenced by Churchill's fears of postwar Soviet intentions. Consistent with the threat of democratic paralysis, he was concerned about the financial collapse of Britain and the need to make it economically and militarily powerful.[33] Yet at the same time Roosevelt opted for a program of "restricted interchange" with the British in order to assure an American nuclear monopoly after the war. The cost of this policy was thought to be at least a six-month delay in the bomb's production.[34]

The president certainly hoped to get along with the Soviets, but his policy, as reflected in his decisions, was to hold the bomb in reserve should efforts at postwar cooperation fail. He never informed the Soviets of the secret, and both he and Churchill rejected Bohr's plans for collaboration with them. In effect, his attitude was one of "carefully guarded skepticism" toward the survival of the alliance.[35] He also appeared to agree with Secretary of War Stimson, who held that

29. Sherwin (1977, 3).
30. On Roosevelt, see Sherwin (1977).
31. Rhodes (1986, 378–79).
32. Cited in Sherwin (1977, 114).
33. Sherwin (1977, 113, 123).
34. Sherwin (1977, 74–76).
35. Sherwin (1977, 8).

the "secret" should be used to gain a quid pro quo from the Soviet Union. In effect, the president may well have been playing a double game with the British and the Soviets, planning to hold out the promise of atomic cooperation with both of them.

Nothing about America's atomic policy was definitely settled when Roosevelt died on 12 April 1945. The new president, Harry Truman, learned from Stimson that the bomb was a weapon of almost unbelievable destructive power with which "even a very powerful unsuspecting nation might be conquered within a very few days by a very much smaller one."[36] His faith in it was quickly stirred by the deterioration in American-Soviet relations in Poland and by those of his advisers who stressed Soviet duplicity.[37] Atomic leverage was thus made the silent partner of economic leverage, particularly in light of the gap between U.S. messianic liberalism and the military realities in Eastern and Central Europe. Even Stimson, who cautioned against getting tough with the Soviets or using the threat of the bomb, termed it "a badly needed 'equalizer.' "[38]

At Potsdam, news of the successful Trinity explosion of 16 July 1945 gave Truman, who had approached his first international conference with trepidation, "an entirely new feeling of confidence." Like the scientists before them, the politicians employed a religious vocabulary to describe their reaction to the bomb. Truman wrote in his diary: "It may be the fire of destruction prophesied in the Euphrates Valley Era after Noah and his fabulous Ark."[39] Of Truman, Churchill noted that "he was a changed man. He told the Russians where they got on and off and generally bossed the whole meeting." Truman was the cowboy with the bomb in his hip pocket. Churchill himself was no less affected: "Stimson, what was gunpowder? Trivial. What was electricity? Meaningless. This atomic bomb is the second coming in wrath." Yet not too much earlier, Churchill had minimized the import of the bomb, thinking that it was just a bigger bang. Now he could argue: "We were in the presence of a new factor in human affairs, and possessed of powers which were irresistible . . . our outlook on the future was transformed."[40]

This faith in an irresistible power was widespread. For Stimson, the bomb was the "master card" in diplomacy as well as "a final arbiter of force." He thought that it would be "decisive" in dealing with the

36. Cited in Fleming (1961, 300).
37. Yergin (1977, 79).
38. Cited in Sherwin (1977, 221).
39. Citations in Wyden (1985, 224).
40. Cited in Alperovitz (1985, 153).

Soviets. The aim was not to intimidate them but to use nuclear weapons as a negotiating tool. Stalin, he believed, could be enticed by the power of atoms. In contrast, James Brynes, who shortly became secretary of state, thought the bomb could be used to dictate terms. Leo Szilard, who met with Brynes in the spring of 1945 to lobby against the planned use of the bomb, reported: "Mr. Brynes did not argue that it was necessary to use the bomb against the cities of Japan in order to win the war. . . . Mr. Brynes' view [was] that our possessing and demonstrating the bomb would make Russia more manageable in Europe. . . ."[41]

That the bomb was a radical and confounding new departure is indicated by the "scientists' revolt" that preceded the bombing of Hiroshima.[42] Where scientists previously disdained any role in politics, they viewed the bomb as something so climacteric that they felt it necessary for themselves, the only ones who had started to grasp its revolutionary character, to make the unprecedented effort to bring politics in line with physics. Where Niels Bohr and Leo Szilard each expressed concerns to top political leaders, scientists at the Chicago Metallurgical Laboratory and other sites circulated petitions and conducted polls over the issue of providing adequate warning to the Japanese. The most enduring legacy of the revolt was the prophetic Frank Report, which noted that nuclear power was fraught with greater dangers than all past inventions. The revolt, which General Groves and Oppenheimer at times sought to suppress, ultimately had little effect, since the information was not passed on to Stimson or Truman. It did, however, set the stage for the scientists' campaign for internationalism in the postwar world.

The politicians, in contrast, were fascinated by this new power and sought means of controlling it and putting it to use diplomatically and militarily. In this context the question usually asked is whether the bombing of Hiroshima and Nagasaki was necessary, that is, could the war have been ended in any other way short of the planned invasion?[43] This query has stimulated considerable research and debate, but it no longer offers new insights. Thus the historians Gerard Clarfield and William Wiecek pose the issue in a different way. They suggest that it may be more useful to ask whether Truman had any reason *not* to use the bomb. Their answer, which is supported by the evidence, is an unequivocal no.[44]

41. Cited in Clark (1980, 218).
42. On the scientists' revolt, see Alperovitz (1985); Gilpin (1962); and Smith (1965).
43. On the bombing of Japan, see Alperovitz (1985); Feis (1966); and Sigal (1988).
44. Clarfield and Wiecek (1984, 67).

Truman, a new president lacking prior knowledge of the bomb as well as international experience, had no contrary evidence about Roosevelt's policies; the advice he received was one-sided; there was a public desire to avenge the "sneak attack" on Pearl Harbor and other "barbaric" acts; the war had already made the indiscriminate bombing of civilians legitimate; the two billion dollars spent on the Manhattan Project needed justification; and, finally, there was the dual possibility of keeping the Soviets out of Asia and gaining some diplomatic advantage from possession of the device.

Careful reading of Clarfield and Wiecek's analysis reveals that use of the bomb was more than a matter of default. Truman had *compelling* reasons to use it. One set of these had to do with keeping the Soviets out of the war, thereby preventing land grabs in Manchuria and their participation in the occupation force. (At Yalta, Roosevelt had negotiated an accord whereby Stalin agreed to enter the war with Japan three months after the defeat of Germany. The Trinity explosion appeared to render unnecessary Soviet participation in the war.)

At least as critical was the issue of control over this extraordinary power. Thus there was the desire "to announce the atomic epic fittingly."[45] Indeed, one can contend that the bombing was in good part an *experiment,* a test to see exactly what we have here, coupled with the need and desire to *display* its worth. There had been much speculation about the nature of this "new factor." But the Trinity success still left great uncertainties. According to Churchill, "No one could yet measure the immediate military consequences of the discovery, and no one has yet measured anything else about it."[46] The bombing was meant to resolve some of the uncertainties. The Americans had to determine what their advantage was worth, and the things that were done (the haste, the secrecy, the selection of targets, the effort to stop Soviet involvement) and the things that were not done (the lack of warning or demonstration, the failure to specify the terms of unconditional surrender or to pursue seriously diplomatic options) made it virtually certain that a military test and display of the bomb occurred.

Unlike the scientists, politicians never seriously questioned using the bomb.[47] After the collapse of Germany, every effort was made to be certain that it was ready in time to use on Japan. The official

45. Fleming (1961, 298).
46. Churchill (1953, 638).
47. See Churchill (1953, 639) and Szasz (1984, 152).

history of the Atomic Energy Commission makes it clear that, early on, the decision had been taken:

> The bomb project had begun as an effort to overcome a Nazi head start. As fears eased that German scientists would win the race, American thinking turned toward Japan. In May 1943, the Military Policy Committee concluded that the Japanese fleet concentrated at Truk would be the best target for the first bomb. Later the same year, Groves approved arrangements to mobilize two B-29s for operations with nuclear weapons. The choice of the B-29 over the British Lancaster, the only other plane sufficiently large, reflected the disposition to use the bomb against Japan.[48]

That prior decision seems to have militated against serious diplomatic efforts to end the war.[49] The ultimatum issued at Potsdam demanded unconditional surrender and warned of "prompt and utter destruction," but did not specifically refer to the bomb. Above all, the ultimatum did not clarify *the* essential point in the demand for unconditional surrender: the disposition of the imperial institution. If the United States, as acting Secretary of State Grew suggested, had guaranteed the continuity of the imperial order, this might well have provided sufficient leverage for the Japanese peace group to prevail over the fanatical militarists.[50] Ironically, the final surrender agreement affirmed the imperial authority of the emperor, subject to the Supreme Command of the Allied Powers.

The Interim Committee and Scientific Panel established to advise Truman on use of the bomb knew, as Oppenheimer put it, "beans" about the military situation and assumed what they had been told: the bomb was needed to avoid an invasion.[51] They spent most of their time discussing a demonstration of the bomb and the question of what to tell the Soviets. Consideration of the Soviets focused on how long it would take them to build a bomb. According to the secretary of the committee, the question of timing "would have a bearing on the subject of how we should try to deal with the subject in relation to the Soviet Union; whether we should try at an early stage to bring about collaboration with them or outdistance them in the race."[52]

48. Hewlett and Anderson (1962, 253).
49. See Sigal (1988).
50. See Powaski (1987, 20–21).
51. Feis (1966, 54).
52. Feis (1966, 40).

Both the aim of keeping the Soviets out of Asia and impressing on them the decisive power of the bomb run through the relevant discussions and remain "unrefuted."[53]

The Interim Committee found many reasons for scuttling the idea of a demonstration blast. The obvious alternative of bombing a thinly populated area of Japan apparently did not arise. Rather, planners had already advised dropping the bomb not only to shock but so as to afford the conditions for an *ideal* field experiment. Five cities had been exempted from conventional bombings in order to avoid confounding effects. The criteria for selecting targets were rigorous. Since the bomb was expected to produce its greatest damage by primary blast effect and its next greatest by fires, the targets were selected to "contain a large percentage of closely built frame buildings and other construction that would be susceptible to damage by blast and fire."[54] Although the bombing was not formally construed as a test, all the elements of an ideal experiment (e.g., naïve subjects, controls, comparison conditions, precise instrumentation) were there. The official aim was to make "the initial use sufficiently spectacular for the importance of the weapon to be internationally recognized when publicity on it is released."[55] Where many scientists thought the spectacle necessary to awaken the world to the necessity of abolishing war, the politicians were more concerned about the future diplomatic and military value of their "winning weapon."

53. Clarfield and Wiecek (1984). James Brynes, soon to be secretary of state, cabled the Chinese leader, Chiang Kai-shek, and asked him to prolong bargaining with Stalin over concessions to be given the Soviets at the end of the war; at Yalta, Soviet entry into the Asian war was made contingent on an accord being reached between Stalin and Chiang. While military commanders believed that the blockade and air raids would end the war with minimal loss of Americans before the planned invasion eleven weeks later (e.g., Szasz [1984, 151]), the American urgency may well have been directed against the Soviets. Blackett (1968, 52) suggests that using the bomb was "not so much the last military act of the Second World War, as the first major operation of the cold diplomatic war with Russia now in progress." Alperovitz (1985) observes that Truman believed that the Soviet declaration of war would be sufficient to cause a Japanese surrender.

54. Cited in Jungk (1958, 177–78).

55. Cited in Sherwin (1985).

4

Managing the Confounding Power

> There is no way to recapture the shock of the bombing of Hiroshima and Nagasaki. It requires an extraordinary feat of historical imagination to recreate the surprise and drama and horror of the day the world first learned of the atomic bomb. And it is all but impossible to recall the instant change in American thinking, the new sense of confidence and power the first atomic explosions engendered.[1]

This is how the historian Gar Alperovitz describes the results of the use of the bomb. As sole possessor of this weapon, America was in a unique position. With the master card shown around the table, there was the hope of winning a substantial pot. The public announcement following the attack on Hiroshima stated that the bomb had more power than twenty thousand tons of TNT. This was, the announce-

---

1. Alperovitz (1985, 188).

ment continued, more than two thousand times the blast power of the British Grand Slam. "It is an atomic bomb. It is a harnessing of the basic power of the universe."[2]

For all the bravado, assimilating the bomb created formidable difficulties for the moral and democratic standing of the United States. On the one hand, it was an extraordinary power that might allow the country to dictate international relations in the postwar world. On the other hand, the bomb was a terrible and destructive weapon that seemed to contradict America's moral exceptionalism.[3] Somehow, possession of this fantastically destructive and secretive device had to be reconciled with the country's special mission and peaceful intentions. Only nations bent on violence could need such a device.

The bomb is probably the best example of what the sociologist James Beckford terms "the power which confounds."[4] Since it transcended the realm of normal comprehension, the bomb produced competing symbolic and practical efforts aimed at managing it. There was, in effect, a contested discourse within the public culture. Different groups drew on one or the other pole of the bomb's numinous ambivalence in order to try to direct the future of the new world. The manifest power of the bomb was confounded by fear, the "hysteria" surrounding radiation and a device that could vaporize cities. Where scientists, some church groups, and a number of journalists played on this fear in order to foster international control, the politicians, while trying to manage their own ambivalence, had to downplay the danger and redirect the terror.[5] So great was the "numinization of politics" that the bomb eventually was rendered safe by symbolically aligning it with elements of American civil religion.[6] In other words, in the process of assimilation, the bomb was invested with a host of symbolic meanings that far transcended the idea of a winning weapon.

2. Cited in Wyden (1985, 288).

3. See Wolfe (1984) for the problems of assimilating the bomb and the emergence of a nuclear folk culture.

4. Beckford (1983). Thus Attlee wrote Truman on 9 August, "the attack on Hiroshima has now demonstrated to the world that a new factor pregnant with immense possibilities for good or evil has come into existence." The prime minister also declared that nuclear weapons were "one of the most difficult and perplexing problems with which statesmen have ever been faced" (cited in Thomas [1987, 441, 451]).

5. See Gamson (1987).

6. The term is drawn from Garret (1974). I will discuss civil religion in more detail below.

Ambivalence is apparent in Truman's reaction to the bomb. Publicly, he called the destruction of Hiroshima the "greatest day in history." He also observed: "What has been done is the greatest achievement of organized science in history. We have spent two billion dollars on the greatest scientific gamble in history—and won."[7] Whereas Truman never ceased to ardently defend his decision in public forums, his private statements reveal a sense of doubt. He justified the bombing as having saved half a million young Americans and because the Japanese had been, in the conduct of the war, "vicious and cruel savages." Yet two weeks before the bombing, he wrote in his diary: "Even if the Japanese are savages, ruthless, merciless and fanatic, we as the leader of the world for the common welfare cannot drop this terrible bomb on the old capital [Kyoto] or the new [Tokyo]."[8] After Nagasaki, he suspended further atomic bombing. Secretary of Commerce Henry Wallace wrote in his diary: "He said the thought of wiping out another 100,000 people was too horrible. He didn't like the idea of killing, as he said, 'all those kids.' "[9] In a cabinet meeting after the London Conference, where Truman worried about the number of divisions he had, Budget Director Harold Smith reminded him that he had the atom bomb up his sleeve. Truman responded: "Yes, but I am not sure it can ever be used again."[10]

The "Great Fear"

The public first heard of the bomb after the shock of Hiroshima. By implication, the sense of the rational and the nonrational, of overwhelming power and nuclear dread, overlapped. Indeed, the latter pole of its numinous ambivalence dominated the initial reception. Thus the historian Ronald Clark suggests that most people "sensed in their bones that the world would never again be the same."[11] Over the next four years, there ensued a symbolic contest between the two poles of the numen. The initial and overwhelming dread was redirected into an air of nuclear omnipotence that resonated with a sense of American uniqueness and goodness.

7. Wyden (1985, 288–89).
8. Cited in Clark (1980, 210).
9. Cited in Herken (1980, 11).
10. Cited in Kurzman (1986, 439).
11. Clark (1980, 128).

Paul Boyer, who has done the most systematic and detailed study of the bomb's initial impact on American thought and culture, presents an array of images consistent with the idea of the numen. According to Boyer, the bomb "bisected history"; from the outset, Hiroshima marked the start of what was called the "atomic age."[12] And what most characterized this new era was fear, or what Boyer terms the "Great Fear." The first NBC news comment following the bombing set the tone: "For all we know, we have created a Frankenstein. We must assume with the passage of only a little time, an improved form of the new weapon we used today can be turned against us."[13] Other descriptions observed that "the whole world gasped" as people were overcome by a "creeping feeling of apprehension," as "one senses the foundations of one's universe trembling." The best-known editorial of the period by Norman Cousins, "Modern Man Is Obsolete" (*Saturday Review of Literature*, 18 August 1945, p. 5), spoke of:

> a primitive fear, the fear of the unknown, the fear of forces man can neither channel nor comprehend. This fear is not new; in its classical form it is the fear of irrational death. But overnight it has become intensified, magnified. It has burst out of the subconscious, into the conscious, filling the mind with primordial apprehensions.[14]

In the "giant shadows" of the mushroom clouds, "all men were pygmies . . . mere foam flecks on the tide."[15] For the public then, the bomb was an "awesome mystery" that unleashed an authentic surge of nuclear fear associated with vaporized cities.

In 1947, David Lilienthal, chairman of the Atomic Energy Commission (AEC), observed that for nearly two years the public had been fed "a publicity diet of almost nothing else but horror stories." He went on to complain that "scaring the daylights out of everyone so no one can think, inducing hysteria and unreasoning fear . . . is not going to get us anywhere . . . we want to go."[16] The scientists, who described themselves as "frightened men," led a concerted campaign of fear, hoping that this would result in a rational political response.

12. See Boyer (1985, chapter 1).
13. Cited in Boyer (1985, 5).
14. Cited in Boyer (1985, 8).
15. Cited in Boyer (1985, 21).
16. Cited in Boyer (1985, 74).

But this was not to be. By harping on the awesome power of the bomb, the scientists left their public "half educated."[17] They were more impressed with the bomb's irresistible power—it was the winning weapon they could rely on—than with the importance of international cooperation as a means of overcoming the terror. The ambivalence that was characteristic of the scientists and the politicians carried over into the public consciousness. But as in the former instances, the sense of power would soon override, at least temporarily, all other considerations.

Public attitudes toward science and the scientists themselves shifted from a sense of awe to one of ambivalence. Initially, the scientists were heroes, the demigods who inaugurated the new world. A *Life* reporter wrote that they had donned the "tunic of supermen" and stood in the spotlight of a thousand suns.[18] Oppenheimer assumed a public stature that was only surpassed by that of Einstein. Physical scientists were in "vogue," wrote a *Harper's* commentator. "No dinner party is a success without at least one physicist to explain . . . the nature of the new age in which we live."[19]

The sense of awe generalized to Trinity. A visit by reporters and photographers in September 1945, undertaken to demonstrate that radiation danger was minimal, gave the site the publicity it had lacked. In the first flush of euphoria, there was the attempt to consecrate it. Numerous proposals were made to turn Trinity into a national monument. In 1965, the Park Service declared it a National Historical Landmark; in 1975, it became a National Historical Site.

But the campaign of fear began to take its toll. Many came to regard the scientists as "wizards of death." Leo Szilard clearly (and perhaps sarcastically) captured the shifting ground: "It is remarkable that all these scientists should be listened to. But mass murders have always commanded the attention of the public, and atomic scientists are no exception to this rule."[20] At the same time, belief in the last stronghold of progress was being undermined. The Calhoun Commission editorialized in 1947 that all hopes for the "redemption of mankind simply through progress in the sciences and technology were permanently wrecked by the latest achievement in that very progress."[21] In a sermon broadcast across the nation, Harry Fosdick complained that the atomic bombing has completed "the most colossal

17. Cited in Boyer (1985, 93).
18. Cited in Kevles (1978, 334).
19. Cited in Kevles (1978, 375–76).
20. Cited in Boyer (1985, 61).
21. Citations in Boyer (1985, 270–73).

breakdown of optimism in history"—the dissolution of the belief in scientific progress.

As noted previously, the Great Collapse was an extended process. The link between science and progress was questioned long before the advent of the bomb. But Hiroshima severed the link more forcefully and completely than anything before. The wedding between science and destruction had been attended by practically the whole planet. Science, when divorced from progress, was reduced to sheer power—to omnipotence. Progress had already enamored us of power. Omnipotence is the orphan of progress; from its forebear it plunders the idea of power, injects it with new content, and elevates it into a singular and overriding principle. But unlike its predecessor, omnipotence admits few moral constraints and offers only feeble conceptions of the good and of human destiny. Despairing of any faith in history, omnipotence cultivates evil. Whether it be cancer or the Soviet Union, the display of transcendent power is facilitated by a worthy, personalized, and demonic opponent.[22]

Civil Religion and the Bomb

Faced with the Great Fear and the special difficulties the bomb posed to American values, politicians needed to redirect or neutralize the

22. Omnipotence also drives us forward in other ways. Specifically, we aim to control and transform practically everything, ranging from disease, to weather, to crops. With genetic engineering, in which we have unlocked "the secret of life," the physical transformation of humans is close at hand. The prospect of uncovering the entire human genetic structure and thereby eliminating maleficent genes adds yet another garland to the human godhead. Other examples of our sense of omnipotence include the "medical miracles" associated with lasers, transplants, and wonder drugs; the development of the computer and "artificial intelligence"; and space travel, with the associated idea of "colonizing" other planets. As compared to these other aspects of omnipotence, nuclearism is unique in several respects: it involves an extraordinary sense of fear; it hinges much more on the creation of an evil opponent; and, at least until now, it portends much greater consequences. In all cases, however, our ability to transform phenomena far outstrips our capacity to either recognize or understand, much less control, the results.

The theology of omnipotence, as it emerges from the ruins of reason and progress, is probably best stated in secular humanism. The idea has even been proposed that humanity—or our superior descendants—will keep going forever. Thus, even as the finite energy of the stars burns out, humanity, through science, will create alternative sources of energy. See C. Lamont (1965).

But since the kind of scientific power postulated in secular humanism was first put on the map by the bomb, the hope in humanism has been perverted by nuclearism. If deterrence theory can be regarded as the theology of nuclearism, omnipotence is reduced to exterminism.

terror. They did so in a number of ways: downplaying the hazards of radiation, the atoms-for-peace campaigns, the focus on the communist threat, and aligning the bomb with American civil religion.[23] Here, I examine first civil religion, then atoms for peace, and finally the communist threat.

Robert Bellah defines civil religion as "that religious dimension found, I think, in the life of every people, through which it interprets its historical experience in the light of transcendent reality."[24] In a pluralistic society, civil religion aims to provide the overarching moral glue. American civil religion is a collection of sacred beliefs that are "formal" and "securely institutionalized." The view that America is God's "New Israel" and that "democratic America has been God's primary agent in history" is integral to the civil religion. For the most part, this faith is invoked on ceremonial occasions: presidential inauguration speeches and national holidays are the best examples. At such times, it is usual to make reference to God, although these invocations are essentially perfunctory. However, use of the civil religion need not involve God. According to one sociologist, there is an increased tendency to make "references to the nation which say it in the place where God had once been."[25] In this more secularized view of civil religion, the nation is "set apart in a special way" and has a special mission. But whether or not God is mentioned, the civil religion functions mainly as a formal source of legitimacy for the constituted political authority.

Bellah observes that the elements of American civil religion were modeled and elaborated in "times of trial."[26] He identifies the War of Independence, the Civil War, and the post-1945 period as times when America tried to find a moral posture in a revolutionary world. It was in the first two crises that the elements of civil religion were forged. In the Civil War, the theme of "death, sacrifice, and birth" is indelibly written into the civil religion. As would be expected, the theme of sacrifice was soon institutionalized and given expression in symbols and rituals, as in Memorial Day.

In the third time of trial, the postwar period, America abandoned (somewhat reluctantly) its isolationism and assumed the role of leader of the "Free World" and global police officer. This new historical

23. Boyer (1985) is an excellent source on both nuclear fear and the various efforts to neutralize or redirect it. See also Hilgartner, Bell, and O'Connor (1982).

24. Bellah (1975, 3). See also Adams (1987); Bellah (1967, 1980); Gehrig (1979, 1981); Hammond (1980); Markoff and Regan (1981); and Marty (1974).

25. Marty (1974, 151).

26. Bellah (1967, 1975).

posture could be seen, *ideally,* as the realization of elements of the country's civil religion. That is, in the early and hence less cynical years of the Cold War, America's "special mission" was readily transferred to the international stage.[27] Here I show that the bomb was assimilated to U.S. civil religion and that it then played a critical role in America adopting the role of Free World leader.

Assimilating the bomb, as noted previously, was problematic. The sense of a disjunction between an open society and a secret and terrible weapon was widely shared. In particular, in the aftermath of Hiroshima there was the fear of humanity having opened up forbidden secrets that it could not handle. Talk of atom smashers in the 1930s and related efforts to split the atom had raised the fear of "penetrating" dangerous secrets. The press and many scientists saw the atom as a repository of the "secrets of the universe" and compared "violent assaults" on the nucleus with more familiar violations of secrets like those committed by Adam and Eve, Prometheus, and Lot's wife.[28] Hence there was a pressing need to demystify the atom.

The numen was initially rendered controllable and legitimate by regarding it as a secret that God had bestowed as a special sign on the American nation. This link between the God of civil religion and the bomb persisted until the Soviet atomic explosion of 1949.[29] Thereafter, references to God were replaced by the idea of "nuclear supremacy," which was tethered to the "special destiny" of the nation. The upshot of these linkages was that a surplus of symbolic meanings was invested in nuclear supremacy.

The fusion of the bomb and the civil religion had a number of significant consequences. First, an affinity was forged between nuclear supremacy and the sanctity of the American way of life. Furthermore, the arms race was moved to center stage in the international conflict

27. Bellah (1975) and Hammond (1980), writing with the hindsight of Vietnam and Watergate, both decry America's failure to live up to its special mission. But before this period, confidence in the nation's mission was still strong, as indicated by Bellah's almost halcyon view of the Kennedy presidency, as compared to the Nixon one.

28. Weart (1988, 55–57).

29. Reference to God in civil religion is not to be equated specifically with the Christian God. There is some irony in the fact that faith in omnipotence was greatest in the United States, which remains the most religious nation among the Western democracies. There is, however, only a loose coupling between Christianity and the bomb. Thus theologians have argued that the bomb's power transcends the powers accorded humanity in the Bible (Kaufman 1985). The churches were among the earliest, and remain among the most persistent, opponents of the arms race. Protestant fundamentalists were also among the isolationists in the early years of the Cold War. Their conversion to internationalism, coupled with faith in U.S. nuclear supremacy, took place in the late 1970s and are strongly linked to the Reagan presidency.

over which of the two rival systems would predominate historically. Third, the panics unleashed by startling Soviet nuclear advances were rendered particularly acute, for they threatened not only national security but the most sacred beliefs about the nation's historic destiny. Finally, the fusion fostered the continued search for the definitive technological solution, for a magical weapon that embodied "invincible omnipotence."[30]

Following the bombing of Nagasaki, Truman began the process of symbolically aligning the bomb with the divine guidance of the nation.[31] According to the president, the atomic bomb was "too dangerous to be loose in a lawless world," and therefore Americans must be "trustees of this new force—to prevent its misuse and to turn it into channels of service to mankind."[32] On 27 October 1945 he announced that "the possession in our hands of this new power of destruction we regard as a sacred trust."[33] Truman thus began to speak of the bomb as a "sacred trust" that had been bestowed on the nation as if by divine right. Secretary of War Stimson spoke of the moral responsibility of America and regarded its trusteeship of the bomb as a "covenant."[34] Even Churchill would reiterate the American claim to a sacred trust when, in his "iron curtain" speech of March 1946 in Fulton, Missouri, he noted that "God had willed" the United States, not "some communist or neo-Fascist state," possession of the bomb.[35]

The sense of a special entitlement imbued with religious metaphors extended to the media and the scientists. *Life* took comfort in the claim that "Prometheus . . . is still an American Citizen."[36] The *New York Journal American* asserted that "Divine providence has made the United States the custodian of the secret of atomic energy as a weapon of war."[37] Arthur Compton, one of the scientists involved in adminis-

30. The term is used by Chernus (1986, 32). See also Benford and Kurtz (1987) and Boyer (1985, 106).

31. There is a certain irony to the American claims, since much of the science and many of the scientists involved in the building of the bomb were of European origin. To a great degree here, the United States, as the last hope of the democracies, was benefiting from the decline of Europe.

32. Cited in Herken (1980, 22). Attlee wrote Truman after Hiroshima and suggested that they should declare their intentions to use "this great power" as "trustees for humanity." See Thomas (1987, 441).

33. Cited in Fleming (1961, 326). See also Pringle and Spigelman (1981, 42).

34. Cited in Herken (1980, 25).

35. Cited in LaFeber (1967, 30).

36. Cited in Boyer (1985, 9).

37. Cited in Boyer (1985, 211).

tering the Manhattan Project, proclaimed, "Atomic power is ours, and who can deny that it was God's will that we should have it?"[38]

Whether it was viewed as a direct bequest of the deity or as a result of the technological might that followed from being the world's greatest democracy, the bomb was both a sign and proof of national grace.[39] Despite the terrible destructive power inherent in the weapon, the American trusteeship would be benign; the bomb would not be used aggressively, but only to preserve freedom. Three months after Hiroshima, Truman said he knew that "because of our love of peace, the thoughtful people of the world know that that [sacred] trust will not be violated, that it will be faithfully executed."[40] A War Department paper of 1947 asserted that Americans were "prevented by our form of government and constitutional processes from launching surprise attacks against potential enemies."[41] In 1949, Vannevar Bush wrote: "We would not strike, because our moral sense as a free people would not allow us to do so; we would not strike because we would sense that we would lose our birthright if we did; but we would also not strike because we would just be cocky enough to believe that we did not need to."[42] The bomb was thus assimilated to America's special moral and historical standing; no other nation was capable of overseeing the trust.

Atoms for Peace

Both the idea of omnipotence and the effort to allay its associated fear are conspicuous in the atoms-for-peace campaigns, which attempted to place a sunny mask on the atom. The promise of an atomic paradise predated the war and was revived in the postwar period. Initially, the benefits of the atom were tied to the idea of international control. Not only would this engender a new type of cooperation among nations, it might end war itself. But with the

38. Cited in Boyer (1985, 212).
39. According to Groueff (1967, 13), "there was something essentially American about the scope of the Manhattan Project. . . . A gigantic scientific experiment on the scale of a whole continent would be performed. It would be the American way of solving the greatest problem of modern science; no other country had the necessary resources and industrial power to attempt it."
40. Cited in Fleming (1961, 326).
41. Cited in Freedman (1981, 36).
42. Bush (1949, 125).

collapse of the effort for international control (to be discussed below), all that was left was the power. Linked to this power, however, were "wonders and portents too numerous to mention."[43] Such marvels as atomic-powered cars and airplanes were promised. Electric energy would be too cheap to meter. There would be enough energy to melt snow as it fell. There was also a plan to melt the polar ice cap with atomic explosions, just one of "the gigantic tasks which man can undertake as an unlimited source of energy now becomes available."[44] Why the ice cap should be melted, given that it could alter the world's climate and submerge coastal cities, seemed to be less the issue than the intoxicating sense of power reflected in the idea.

In its peaceful incarnation, the atom held the promise of unlimited energy to fuel economies that had been stalled by the depression. The new British prime minister, Clement Attlee, in his first public declaration about the bomb, spoke of atomic energy as a possibly "perennial fountain of prosperity."[45] Henry Wallace drew on the winning-weapon theme when he voiced the widely held view that atomic energy could undo the economic *and* political crises threatening the West: "Communism can only thrive on scarcity and inequality. Atomic energy properly applied can mean the disappearance of both."[46] Such ideas were soon embodied in the "atoms-for-peace" campaigns.

The first incarnation in the late 1940s promised much and delivered little. However, the idea was revived by the Eisenhower administration in the mid-1950s. Consider how the AEC chairman, Lewis L. Strauss, celebrated the first power fed into a utility grid from a nuclear reactor in 1955:

> Before me stands a large, two-way switch. . . . If I throw the blade in one direction, it will turn the propeller shaft of a military weapon. But when I throw it in the other direction, as I am about to do, it will send atomic electric power through transmission lines to towns and villages, farms and factories— power not to burst bombs or propel submarines, but to make life easier, healthier, and more abundant. The switch is a symbol of the great dilemma of our times. I throw it now to the side of the peaceful atom. . . .[47]

43. Cited in Boyer (1985, 110).
44. Cited in Boyer (1985, 111).
45. Cited in Thomas (1987, 441). The speech was drafted beforehand by Churchill.
46. Cited in Walton (1976, 124).
47. Cited in Hilgartner, Bell, and O'Connor (1982, 44).

From the present perspective, where Three Mile Island and Chernobyl are everyday terms, the atomic utopia no longer resonates positively. What remains clear, however, is that these promises were made without a trace of evidence to substantiate them. They were made because there was a pressing need to "denature" the atom, to soothe public fears and to equate atomic research with progress. Atoms for peace was a feeble attempt to fortify these links and to domesticate the atom. The campaign was a short-lived fiasco. But without a progressive underpinning, the atom is reduced to a weapon. The only benefit it brings is to hold Satan at bay. Its dark side was to replace its sunny countenance.

The Bomb and the Cold War

This is what happened. The Great Fear fostered by Hiroshima, the fear of irrational death, of mass death, of the social death, was initially a generalized and abstract fear, an unaffiliated terror. It envisioned future developments, but was not directed at a specific enemy. When the abstract fear intersected with the Cold War and the absolute enemy, Americans embraced the bomb. It became the solution, and Americans were now ready to accept practically any measure that promised to sustain their nuclear supremacy; terror was to be held at bay by augmenting the terror.

This book focuses on the arms race, and it cannot make any pretense to undertake a history of the Cold War. But the claim that the bomb was received as a numinous and confounding phenomenon does suggest that it played a greater part in initiating the Cold War than is usually supposed. Most histories of the Cold War discuss the bomb in considerable detail but do not assign it a clear role; this is probably consistent with the diplomatic focus of such works and the fact that the bomb was rarely brandished openly.[48] But some writers do assign it a more specific role. David Horowitz, for example, suggests that the worldwide network of bases from which the United States could launch a nuclear attack on the USSR was a critical determinant of Soviet behavior in the Cold War.[49] Herken claims that the bomb resulted in the militarization of the Cold War.[50] Boyer

48. Both Thomas (1987) and Yergin (1977) are good examples of this.
49. Horowitz (1965).
50. Herken (1980).

suggests that fear of communism was used to redirect fear of the bomb.[51] For our purposes, Boyer's analysis is the most critical. But in contrast with him (and more consistent with Herken), I suggest that the fear of communism did not redirect nuclear fear but that the two fears intersected, thereby producing what might well be the greatest cleavage in history.

Like Boyer, I am broadly concerned with how the bomb affected American attitudes toward the Cold War. The most important question has to do with the "selling" of the Truman Doctrine. In early 1947 Americans had yet to embrace Cold War attitudes and still held some admiration for their former ally for the role it played in the defeat of Nazi Germany.[52] Without a genuine conversion of attitudes toward Russia or America's role in international politics, and in the absence of an international incident, how did Truman sell a decision that was "more serious than had confronted any President?"[53] More specifically, how did Truman persuade a population that was still war weary and uncertain about the nation's international role to accept an ideological program that many regarded as "giveaways" and an "intervention" in remote places with corrupt and undemocratic governments? The usual explanation asserts that he sought, as Senator Arthur Vandenberg put it, "to scare the hell out of the country" in order to garner support for this radical shift in doctrine.[54] In contrast, I suggest that Truman's success had at least as much to do with the Great Fear, atomic spies, and the failure of talks on international control than with any other factor, including the reservoir of domestic anticommunism, Soviet behavior in Eastern Europe, and the administration's scare campaign.

An analysis that relies on "electrifying" speeches cannot be sustained. Research on moral panics has focused on how the media and public figures mobilize popular opinion by exaggerating and distorting threats. Yet as that research makes clear, social reaction is not enough: moral panics *must* draw on and evoke deeper or latent

51. Boyer (1985, 102–3).
52. On American attitudes, see Yergin (1977, 275–302). He shows that whereas the anticommunist die was strong among the elite, public attitudes were still contradictory and wavering. The best discussion of the whole issue is provided by Jones (1955). His description of the obstacles to the acceptance of the Truman Doctrine is so thorough as to make his assertion that "within three days [of Truman's address to the nation] it was reasonably clear that Congress and the American people would approve the course of action recommended by the President" (1955, 173) something of a surprise. Jones talks of a complete and genuine conversion without specifying a conversion process.
53. Cited in Jones (1955, 39).
54. See Sanders (1983, 15) and Yergin (1977, 275–302).

tensions and fears.[55] Without such fears to catalyze the people, leaders can in effect be hostage to public opinion and sentiment. Thus both before and at the very end of World War II, Roosevelt felt stymied by public attitudes and resistance. Whereas Truman and Eisenhower had little use for McCarthyist smear campaigns, both remained publicly circumspect. But an analysis focusing on McCarthy's speeches and ignoring the latent fears he addressed (the Soviet atom bomb, the fall of China) will not suffice.

A further example, which anticipates the analysis in Chapter 5, is pertinent. According to Fred Knelman's study of the arms race, there were eighteen ostensible "gaps" in favor of the Soviets (the actual number is substantially higher, since some of them were renewed and/or intensified several times).[56] These gaps were often fabricated, and they were invariably promoted by elements of the military-industrial complex (MIC) with the aim of peddling fear in order to provide a pretext for funding new nuclear buildups. But despite the fact that the Soviet nuclear menace was invoked so frequently in the postwar period, only three of these gaps gave rise to either panics or large buildups. The others were either met with apathy or actually backfired on their proponents. That most gaps did not produce panics or buildups implies that the stoking of the Great Fear is neither automatic nor easy. In other words, the populace is not just a mass ready to be swayed by its leaders.

According to Jerry Sanders, Truman successfully oversold the foreign threat of communism by drawing on the reservoir of *domestic* anticommunism.[57] There is certainly something to this claim, but it leaves problematic the link between internal communist subversion and the situation in Greece. For most Americans, the latter was too distant to really matter, especially in a context where the global definition of American national security had yet to take hold (the Truman Doctrine was itself a powerful assertion of that definition). Here I suggest that what forged the links among these factors was the bomb. Where Boyer proposes that a public in the grip of atomic fear was encouraged to redirect its anxiety ("Not vaporization but communization was the great menace confronting mankind"),[58] I contend that the two fears intersected and forged the dynamic between nuclear omnipotence and nuclear panics.

55. Cohen (1972, xxiv, 193).
56. Knelman (1985, 122).
57. Sanders (1983, 16–18).
58. Boyer (1985, 102–3).

The link between the confounding power of the bomb and political reality was forcefully drawn in notes by Secretary of War Stimson:

> S.1 [the atomic bomb]
> Its *size* and *character*
> We don't think it *mere* new *weapon*
> *Revolutionary Discovery* of Relation of man to universe
> Great History Landmark like
>> *Gravitation*
>> *Copernican* Theory
> But,
> Bids fair [to be] *infinitely greater,* in *respect* to its *Effect*
> —on the ordinary affairs of man's life.
> May *destroy* or *perfect* International Civilization
> May [be] *Frankenstein* or means for World Peace[59]

Stimson told Truman in September 1945, "I consider the problem of our satisfactory relations with Russia as not merely connected with but as virtually dominated by the problem of the atomic bomb."[60] Here I propose that the role of the bomb in Soviet-American relations devolved in particular around one element of its seeming omnipotence, the idea of an "atomic secret."

The belief that the bomb was a sacred trust implied two things: that there was an "atomic secret" and, by implication, that the United States would have an enduring atomic monopoly. Consistent with the idea of divine entitlement, there was widespread belief in an atomic secret based on something like a magical formula written on a single sheet of paper.[61] Hold the formula tightly, and the United States could have an indefinite monopoly on atomic bombs.

Such was not to be the case. Soviet challenges to the American sense of nuclear omnipotence were to create moral panics that battered the country's view of its own historic destiny as embodied in its civil religion. But until the Soviet atomic explosion of August 1949, faith in an atomic secret was strong, and central to the U.S. posture in the Cold War. The idea of a secret was denied by scientists but believed by much of the public and the policy elite, with the result that it influenced attitudes, diplomacy, and military policy. The atomic secret and its attendant monopoly were critical in a number of ways:

59. Cited in Rhodes (1986, 642); italics in original.
60. Cited in Herken (1980, 23).
61. See Herken (1980, 127).

they opened the doors to the "atomic diplomacy" that characterized immediate postwar relations between the two powers; they gave rise to the fear of atomic spies and Soviet duplicity; they authorized the United Nations debate on international control; and they afforded a sense of confidence, a belief in an asset that probably fostered U.S. involvement in international affairs. While counterfactual "ifs" are inherently problematic, both U.S. confidence and involvement in international affairs would probably have been significantly diminished or taken much more time to develop without the device.

In diplomacy, the nuclear monopoly was expected to work like a magic wand. James Byrnes, who became secretary of state in 1945, conjectured at Potsdam that "the atomic bomb assured ultimate success in negotiations."[62] But at the London Conference of 1945, the first postwar meeting of the big-three powers, Byrnes's effort to brandish the "bomb on his hip pocket" was a fiasco. No one had expected a failure in the first postwar conference; but instead of making the Soviets more tractable, the bomb rendered them more intransigent. It was creating rancor and distrust, and Byrnes would later call it "the cause of all his troubles."

Sensing the futility of atomic diplomacy, he reversed his strategy at the Moscow Conference and reached a tentative agreement for atomic collaboration. But his success was undermined at home by a broad front of opposition that included the president. Senator Vandenberg termed Byrnes's plan "one more typical American give-away."[63] Sentiment against sharing the secret was "enormous," with more than 90 percent of the Congress and 75 percent of the public opposed. Other polls appeared to contradict this, revealing a high level of generalized support for international control and world government.[64] But support for these ideas evaporated when the questions were *specific*. Americans were not yet ready to cede any national sovereignty or perceived advantage to others. More broadly, the contradictory results appear to reflect the two poles of the bomb's numinous ambivalence: extraordinary power and extraordinary dread.

In September 1945 Prime Minister King informed Truman that a Russian spy ring was operating in Canada. The president advised against premature action, since he knew that there was "no precious secret" and had put his faith in America's industrial capacity and

---

62. Cited in Thomas (1987, 442).
63. Cited in Yergin (1977, 153).
64. See Boyer (1985, 37–38, 56–57) and Herken (1980, 32).

resources as the ground for monopoly.[65] Up to this time Truman had debated whether to make an open approach to the Soviets about the bomb. According to Hugh Thomas, the president's decision not to make such a move "was probably affected" by the news of Soviet spying.

On 3 February 1946, Drew Pearson told the nation on his radio show that a Soviet spy ring was operating in Canada. No details were given. In the next few days, the spy ring came to dominate the news. At an 8 February news conference, Byrnes was unable to diminish the concern. "Atom spies" had become a cause célèbre. The American publicity forced the Canadians to arrest twenty-two suspects before their investigation was complete. On the same day, Washington columnist Frank McNaughton disclosed from a "confidential source" that the Russian spies in Canada were after atomic scientists and information, and that a second ring was active in America. According to the informant, the FBI would have moved to break up this second ring "but for the arguments of state department men—who [the source] will not name—that to do so would upset our relations with Russia."[66] For a population already in the grasp of the Great Fear produced by Hiroshima and the scientists' crusade, the spy scandal produced a near hysteria.

As it turned out, the atomic spy ring in Canada was unspectacular and after essentially insignificant information. Open-market purchases better characterized it than cloak-and-dagger operations. On the whole, it was as much an atomic-spy hoax as an actual ring (especially in the United States proper). Yet it did foreshadow future spy cases, real and imagined. While Truman privately downplayed the scandal, he remained publicly circumspect. In all probability, he was both surprised and cowed by the strength of public reaction.

The scandal rapidly eroded support for the McMahon Bill, which was aimed at establishing civilian control over atomic energy in the United States. As one newspaper put it, the "mood for clutching the secret tightly, rather than letting out a single phrase of it, appeared wholly predominant."[67] The bill did pass, but with several extremely significant amendments that reversed much of its original intent. Specifically, it established a military liaison committee that could

65. Cited in Herken (1980, 127). Mackenzie King, on the other hand, did believe in the secret: "So far as Canada was concerned, however, it was not correct to say that we had the secret of the bomb because we had not had to do with its manufacture at all" (cited in Hyde [1980, 38]).

66. Cited in Herken (1980, 129–30).

67. Cited in Herken (1980, 127).

review and appeal the decisions of the civilian commissioners directly to the president. The army would also continue to control the stockpiling of fissionable materials and retain authority for collecting and analyzing intelligence on foreign efforts to build the bomb. In addition, where the original version of the bill stressed the free dissemination of information, the focus was now on restrictions. The bill effectively barred America's allies from all atomic information.

The spy scandal paralleled a number of other developments that exacerbated relations between the Soviets and the Americans. In the Iranian crisis, Truman sent a battleship to Turkey after Stalin delayed the removal of his troops from northern Iran. On 22 February 1946 George Kennan sent his famous "long telegram," which fully subverted the notion that Russia could be managed by regular big-power politics. This was followed by Churchill's "iron curtain" speech.

It was in this context that Bernard Baruch led the final effort at international control at the United Nations. Of Baruch's appointment, Byrnes was to say, "This is the worst mistake I have ever made, but we cannot fire him now." Yet the Senate did not regard it as a mistake and confirmed Baruch without calling him to testify before the Foreign Relations Committee. The wide acclaim given the appointment underscores the dilemma faced by the administration. Because of the widespread belief in the atomic secret and the reluctance to give anything away, the administration had to appease domestic critics both in and out of Congress. But what was acceptable at home was not acceptable to the Soviets. Baruch used the U.N. platform to win the propaganda game at home.[68] The Soviet rejection of the plan, following in the wake of the spy scare, was taken as a sure sign of their aggressive intentions. For some, the refusal was a prelude to hysteria. Senator McMahon publicly spoke of a preventive war against the Soviets, observing "that for the first time in human history, the failure to agree to a sane, effective and righteous control of weapons of war constitutes in and of itself an act of aggression."[69]

Consistent with this analysis, it was then that calls for "preventive war" became common. They came from General Groves, Senator McMahon, Bertrand Russell, Winston Churchill, Leo Szilard, George Kennan, and a variety of generals.[70] Such was the power attributed to the bomb that it could not only win a quid pro quo from the Soviets but it could also be used to drive out the rulers of the Kremlin and

68. Herken (1980, 177).
69. Cited in Clarfield and Wiecek (1984, 97–98).
70. Trachtenberg (1988–89, 7–11).

free Eastern Europe from Soviet domination. The notion of a "justi-fiable war" was stated admirably by Bertrand Russell:

> There is one thing and only one thing which could save the world, and that is a thing which I should not dream of advocat-ing. It is, that America should make war on Russia during the next two years, and establish a world empire by means of the atomic bomb. This will not be done.[71]

In this regard, the physicist P.M.S. Blackett wrote, "When a nation pledges its safety to an absolute weapon, it becomes emotionally essential to believe in an absolute enemy."[72] This is precisely what happened. When the Great Fear intersected with atomic spies, and the Soviets rejected the seemingly benevolent American plan for international control and asserted their dominance in Eastern Eu-rope, the extraordinary weapon found an extraordinary foe. The Soviets (and I will discuss this in more detail in Chapter 8) were rendered satanic, and this in more than a loose or metaphorical way. The powers attributed to them reproduced many of the more venal powers attributed to Satan.

Americans had put their faith (and fate) in a numinious power and the secret of its creation. In the struggle between the forces of good and evil, the numinous power had gone to the good. The new world was their world alone. It was an exalting and terrifying world. Yet a satanic force was using deceptive means to steal the secret. Loss of the secret was the most terrifying of all possibilities. For it not only meant that the power had to be shared (as with Britain, for example), but that the force would be available to such an evil enemy. To an extent, the Soviets became the externalization of the terror Americans felt at their own numen. The possibility of the awe-inspiring yet diabolical weapon falling into the hands of a diabolical foe engendered hysteria.

A central feature of the powers attributed to the Soviet devil is subversion—the power to infiltrate, to use deception and perfidy in order to become, in a recent phrase, the "engineer of human souls."[73] In considering the role of the bomb in the formation of Cold War attitudes, it is important to recall how much these attitudes were concerned with secrecy and security. Thus, shortly after he an-nounced the Truman Doctrine, the president put security on the map

71. Cited in Clark (1980, 241).
72. Cited in Moss (1968, 109).
73. This is the title of an excellent novel by Josef Skvorecký (1988).

by establishing the Federal Loyalty and Security Program to vet federal employees.[74] The House Committee on Un-American Activities also began to investigate the Manhattan Project and atomic scientists. Over the next few years the concern with secrecy and security would be the commanding concern of an America caught up in the first postwar moral panic. Here again the atomic bomb and atom spies play a critical role in the creation of the public culture surrounding the Cold War.

74. Truman put the loyalty boards in place as a response to domestic political pressures: he wanted to sap the thrust of Republican anticommunism and the possibility of Congress passing even more severe security measures. But I am suggesting that the underlying source of these domestic pressures, what made them emotionally and politically compelling, was the threat of Soviet spies, particularly atomic spies.

5

The Soviet Atomic Bomb–
Korean War Panic

According to Paul Boyer, "The process that had begun gradually around 1947, in which the image of the bomb as a menace to be eliminated was effaced by its image as a vital asset in the intensifying struggle with the Soviet Union, vastly accelerated after September 1949."[1] It was in that month that Americans first learned that the Soviets had exploded an atomic device. Now the "dread destroyer" of cities was metamorphosed into "the shield of the Republic." Here I argue that the loss of the atomic monopoly was a critical factor in initiating the first postwar moral panic. That panic commenced with National Security Council Memorandum 68 (NCS-68), provided a powerful impetus for an arms race infused with practical and symbolic significance, and culminated in a frantic McCarthyist hunt for traitors.

---

1. Boyer (1985, 339).

Virtually every attempt to explain the arms race draws, at some point, on the fear of the Soviet threat.[2] But for the most part this fear is treated as a primitive concept that is not systematically developed, analyzed, and applied. Above all, fear is treated as a constant: C. A. McClelland, who is typical, uses the label "Cold War pattern" to characterize the period from 1948 to 1964 as one of the highest sustained tension.[3] I suggest that the fear fluctuates, and that it does so in ways that are consistent with fluctuations in the arms race.

To sharpen and focus the analysis of fear, I approach the arms race through the concept of moral panic. I argue that the generalized and relatively constant fear of the Soviet threat was punctuated by moral panics unleashed by the perception of spectacular and startling Soviet challenges to American *nuclear* hegemony. These spectacular developments stoked the "Great Fear," caused sudden and dramatic shifts in American attitudes, and resulted in massive nuclear buildups. I identify four panics: the Soviet atomic bomb–Korean War panic, *Sputnik*, the Cuban missile crisis, and the window of vulnerability–Afghanistan crisis. *Sputnik*, which produced the bogus "missile gap," and the Cuban missile crisis were "pure" nuclear panics, portending a direct threat of vaporized cities. The first and last were "mixed" panics, involving significant nuclear threats coupled with political and military elements.[4] The growth of the American arsenal came in three large waves (the Truman, the Eisenhower/Kennedy/McNamara, and the Reagan buildups), each preceded by a moral panic (the Soviet atomic bomb, the *Sputnik* and Cuban missile crises, and the window of vulnerability, respectively) as an immediate determinant of the buildup.

A major goal of my analysis is to document this relationship between the panics and the buildups. The analysis also aims to show that the panics greatly exaggerated the reality of the Soviet threat. Of course, since the ending of the Cold War and the revelations made available through *glasnost*, it is abundantly clear just how much Americans exaggerated the threat. While the information available when the panics occurred indicated great exaggeration at the time, this amplification of the threat is quite understandable in the context of nuclear fear.

2. See Kurtz (1988, 180–88).
3. McClelland (1977); see also Abolfathi (1980).
4. Indeed, the last panic did not actually involve a startling nuclear threat. Rather, it was set in motion by the perceived affinity between Soviet adventurism (particularly the invasion of Afghanistan) and their possession of a nuclear advantage (the ostensible window of vulnerability).

A secondary goal is to show that the pattern of "panic and reaction" also helps to explain the ascendency of the military-industrial complex (MIC) in the United States.[5] Many analyses of the arms race focus on the role of the "iron triangle" in manipulating fear of the Soviets and in institutionalizing a spiraling arms economy so intertwined with the domestic economy that the link is practically irreversible.[6] Yet Paul Koistinen's historical survey of the power of the U.S. military reveals that it has been "derivative" and "secondary." If the United States has developed a "permanent war economy," it is a recent phenomenon that makes it difficult to claim that it is endemic to capitalism.

More important, Koistinen suggests that the MIC does not appear to have *originated* the conditions necessary for its own ascendency.[7] Following this line of reasoning, I contend here that efforts by the MIC to create outbursts of fear that were sufficiently dramatic to provide the political leverage for large nuclear buildups (as opposed to maintaining budgets or programs) were of limited success. Of the more than eighteen attempts to create fear over gaps favoring the Soviets, only three unleashed the panics associated with massive buildups (the Cuban missile crisis did not involve a gap but a sudden shift in the deployment of nuclear forces).[8] Among the key factors differentiating the three successes from the much larger number of failures were spectacular Soviet nuclear advances that unleashed authentic outbursts of fear. In other words, the *timing* of the first two buildups was determined by surges of fear that, in the context of so many failed efforts at fear creation, must be regarded as more authentic than manipulated.

However, the perception of spectacular Soviet nuclear advances was a necessary but not sufficient condition for creating panic. The surges of fear that these advances engendered were largely inchoate; it was incumbent on elite groups to impart *direction* to them. In particular, the MIC dramatized the fear by creating an affinity between nuclear supremacy and both national security and the sanctity of the American way of life. The sociological concept of "elective affinities" suggests that actors' choice of actions are given by their universe of meanings.[9] In the postwar United States, the concept of national security was so thoroughly militarized that the idea of nuclear build-

5. On panic and reaction, see McDougall (1985, 422) and Ungar (1990a).
6. On the MIC, see, for example, Koistinen (1980); Melman (1970, 1974); Mills (1959); Moskos (1973); Prins (1982); and Kurtz (1988).
7. Koistinen (1980, 123).
8. On the gaps, see Knelman (1985, 122).
9. On elective affinities, see Howe (1978).

ups to counter any and all perceived threats virtually monopolized public discourse and decision making.[10] The militarization of national security simply equated it with an American nuclear monopoly or lead. So potent was this affinity that it practically banished other models of security (political agreements, arms control) to the realm of propaganda. At the same time, an affinity was forged between (technological) superiority in the arms race and the other ways in which the United States presented itself as the superior model for the rest of humankind.[11] That is, the idea of America's special destiny as embodied in its civil religion became strongly linked with its nuclear superiority. The forging and dramatization of these affinities meant that spectacular Soviet nuclear advances engendered unprecedented levels of insecurity and uncertainty, which were then mitigated by corresponding American nuclear buildups. In effect, moral panics led to titanic efforts by America to reassert its sense of nuclear omnipotence.

This chapter first outlines the concept of moral panic and applies it to the international system. Following a discussion of the paradox of indispensability and the power of the military after World War II, I discuss the first panic by examining the Soviet nuclear spectacular that affected its timing, the affinities forged between nuclear supremacy and both national security and the sanctity of the American way of life, and the actual consequences of the panic for arms buildups.

Moral Panics and the Anarchical International System

Stanley Cohen's definition of moral panics has become the standard:

> Societies appear to be subject, every now and then, to periods of moral panic. A condition, episode, person or groups of persons emerges to become defined as a threat to societal values and interests; its nature is presented in a stylized and stereotypical fashion by the mass media; the moral barricades are manned by editors, bishops, politicians and other right-

10. On the militarization of national security, see Yergin (1977, 196).
11. McDougall (1985, 227).

thinking people; socially accredited experts pronounce their diagnoses and solutions; ways of coping are evolved (or more often) resorted to; the condition then disappears, submerges or deteriorates and becomes more visible.[12]

The initial work on moral panics, which emerged in the study of deviance, focused on how moral entrepreneurs create or amplify panics for their symbolic value: conflicts are used by one moral universe to impose its values on a different and ostensibly inferior moral universe. More recent work adds an interest dimension to the concept, showing that elites can use moral issues to cloak struggles between different groups with competing interests.[13] Nachman Ben-Yehuda suggests a synthesis of the two approaches whereby moral and interest factors operate simultaneously.[14]

Here I apply the concept to international affairs, in particular to the anxieties produced by nuclear weapons. It has long been recognized that the social construction or exaggeration of an external threat is an excellent substitute for the traditional alliance of throne and altar as a basis for social order.[15] The "folk devils" investigated by moral-panic researchers have included various youth movements, as well as witches and religious cults. The idea of folk devils certainly applies to the fear of the Soviet Union and of internal communist subversion.

12. Cohen (1972, 9).
13. Galliher and Cross (1983).
14. Ben-Yehuda (1986). See also Ben-Yehuda (1980, 1985); Cohen (1972); and Ungar (1990a). A more historical approach can be found in Hofstadter (1967) and Levin (1971). The discussion in the text avoids technical exposition. Sociologically, moral panic is identified with two things, exaggeration of the actual threat and the process by which elites construct or amplify panics. Exaggeration, however, is an extremely controversial issue in the study of moral panics. Waddington (1986), who provides the best discussion of what he terms the problem of "disproportionality," concludes that the study of moral panic is often ideological, since researchers usually do not provide sufficient evidence that the scale of response is significantly greater than the scale of the problem. In the panics studied here, I think it quite clear that Americans exaggerated the *immediacy* of the Soviet threat. *Sputnik* did not give rise to a missile gap on the Soviet side; indeed, the opposite was true. But it did portend future Soviet missile deployments that left the United States vulnerable to exterminism; hence the panic. Applying the concept to the arms race leads to the following specific differences from its "conventional" use in the study of deviance: first, in the arms race, there are authentic outbursts of fear that appear to be more extreme than the fear associated with deviance-based panics; and second, in the arms race, elites (or the MIC) appear to have considerable difficulty fabricating panics and serve more to impart direction than to create them.
15. McNeill (1982, 379–80).

The anarchical international system has some immediate theoretical implications for conceptualizing moral panics. These can be grasped by relating them to the distinction between the *timing* and the *direction* of panics.[16] The former focuses on the question of why the panics happened when they did, while the latter is concerned with the specific content of the panics, particularly the solutions developed for them.

In the international realm, the episodes or groups that pose a threat to societal values and interests are essentially foreign. It is a conflict between nation-states and their agents that is in question. Unlike the panics studied from a labeling perspective, international moral panics do not necessarily evolve around specific subpopulations becoming deviant.[17] The variables that affect the possibility of rendering subpopulations deviant—the ability to mobilize power, the perceived threat potential in the issue, public awareness, and the resistance encountered—do not operate in the same way in the international system. By implication, the international context affords less control than the domestic one over the timing of the precipitating events (nuclear advances, invasions) that provide the dramatic basis for panic. In short, elites are neither totally in charge of nor do they monopolize the social construction of fear.

At the same time, international moral panics contain the potential for the most extreme manifestation of panic whose direction is under a high level of domestic (and hence elite) control. First, these panics superimpose national differences on the threatening development; hence, symbolic values *and* interests will inevitably overlap. Second, in the context of the anarchical international system, it is far easier to vilify the offender than to bring any other mechanism of social control to bear. The inability to directly redress threats that often implicate sacred national symbols can unleash *ritual* responses, domestic moral crusades that have little bearing on the threat but dramatize insecurity and prime the populace for titanic struggles.[18] National symbols, values, and fears are selected and dramatically employed by interested elite groups to create elective affinities that authorize solutions to the perceived threat consonant with their own interests.

Our primary concern is with the timing of panics. The aim is to show that spectacular Soviet challenges to the American sense of nuclear hegemony touched off moral panics that afforded the politi-

16. See Ben-Yehuda (1986).
17. Schur (1980).
18. See Kurtz (1988, 71–75). McCarthyism is, of course, the prime example.

cal leeway for large-scale nuclear buildups. Where public opinion is usually an amorphous phenomenon that does not readily influence the political process, in the context of panic that opinion is mobilized through the media by various elites and moral entrepreneurs. That is (and unlike the generalized fear of the Soviets), nuclear panics entail such startling threats to societal values and interests, and occasion such extreme "emotion work," that there is usually concerted pressure among the elite *and* the public for actions to redress the threat.[19] Nuclear panics are so extreme because of their latent threat potential: the fear of vaporized cities, irrational death, mass death, the social death—the Great Fear. Indeed, because of the bomb's numinous nature, nuclear panics can well be considered panics *of* omnipotence. They simultaneously threaten the historical continuity and destiny of the nation in symbolic and practical ways.

Because of its very extremity, the stoking of the Great Fear is never easy or assured. Overall, people prefer to avoid thinking about the nuclear threat and tend to respond to it only when it is forcefully impressed on them. According to Robert Divine, for example, the fearful effect of the H-bomb was too great to contemplate and public fears were focused instead on the more manageable problem of radiation poisoning. The fallout scare of 1959 thus produced an "alarm bordering on panic," but not a full-blown panic.[20] More broadly, many other factors appear to affect the level of nuclear fear. Psychic numbing, the loss of immediacy, perceptions of decreased risks, deterrence theory (as opposed to the theory of fighting a nuclear war), and the neutralizing effects of the peaceful atom are among the elements that diminish the fear. In contrast, an emphasis on civil defense, atmospheric nuclear testing, superpower confrontations, and Soviet nuclear advances all stoke the fear.[21]

When viewed against the preference to avoid thinking about the nuclear threat, even the much-vaunted ability of the MIC to create fear must be regarded with suspicion. As noted previously, only three of the many "gaps" peddled by the MIC were effective, the remainder either being met with apathy or actually backfiring on their proponents. That most gaps did not produce panics or buildups is something that needs to be taken into account by theories that purport to explain the dynamics of the arms race.

I suggest that startling Soviet nuclear advances (the atomic bomb

19. On emotion work, see Hochschild (1979). See also Edelman (1971).
20. Divine (1978).
21. On nuclear fear, see Blackett (1949); Boyer (1980, 1985); Divine (1978); Mandelbaum (1981); and Weart (1988).

and *Sputnik*) or what appear or are construed to be dramatic shifts in the nuclear balance (Cuban missiles, window of vulnerability) are required to stoke the Great Fear. That is, the Soviet nuclear advance must be sufficiently spectacular to render visions of vaporized American cities palpable *and* simultaneously to challenge the belief in America's special historical destiny. This engenders an inchoate sense of panic, one that remains to be dramatized and thereby directed. All panics require direction, often under conditions of elite conflict. While part of the elite may seek to suppress the panic, other elite elements will attempt to mobilize fear in particular directions and use public sentiments to "bludgeon" opposing elites.[22]

In full-blown nuclear panics, however, the fear is so extreme that efforts at panic suppression are quickly shunted aside by the semiautomatic consensus for the reassertion of American nuclear supremacy. The affinities associated with nuclear supremacy create the perception that both national security and the sanctity of the American way of life are imperiled. Since superiority in the arms race comes to be associated with the very superiority of the capitalist-democratic system in its worldwide struggle against communism, these panics foster titanic efforts to reassert American nuclear—and moral—hegemony.

Nuclear moral panics, as opposed to the generalized fear of the Soviets, are in fact comparable to the apocalyptic movements of earlier centuries. Spasmodic outbreaks of religious fear and fervor, associated with the plague, earthquakes, and various prophecies, have punctuated Christian history.[23] Despite the efforts of the church to discredit and suppress it, the apocalyptic tradition persisted in the obscure underworld of popular religion. The nuclear world is also punctuated by outbursts of fear. Efforts to discredit or suppress the fear have also been unavailing. Among works of modern astrology and mysticism, perhaps nothing stands out so potently as Hal Lindsey's association of the bomb with the eschatology of the Revelation of Saint John the Divine. His book, *The Late Great Planet Earth,* has sold more than twenty million copies, and its prophecy of Armageddon has been echoed by President Reagan.[24] And for the large number of people who tend to regard Revelation as a spiritual allegory, the relentless pursuit of the divine power of exterminism means that the fear of God has been supplanted by the fear of humanity—of Armageddon without the Second Coming. Just below

22. Sanders (1983, 85–115).
23. See Cohn (1970).
24. On Reagan's apocalyptic pronouncements, see Halsell (1986).

the surface of the nuclear world lie extraordinary anxieties that are difficult to bring to consciousness but that, on occasion, erupt in violent convulsions.

The Paradox of Indispensability

Closely linked to the view that the arms race was essentially a panic-induced phenomenon is what can be termed the *paradox of indispensability*. On the one hand, the bomb is not an ordinary weapon; in fact, it is not a "weapon" at all. Or, to put it another way, the bomb is not *best* understood as a weapon. Rather, it is a source of numinous terror and an instrument of mass death, if not extinction. Discussion of the status of nuclear arsenals—whether they are "indefensible," for instance—will be threaded throughout this and subsequent chapters. Those discussions aim to show that the arsenals are *prostrate*— that they are essentially useless and, by implication, have a strong paralytic effect on their possessors. Above all, I will contend that these weapons are *politically* prostrate; for the most part, their use is not a real option for political leaders.[25]

On the other hand, both before and certainly ever since the Trinity explosion, the bomb has been considered *indispensable*. More and more powerful bombs, allied with more and more sophisticated delivery systems, have been assembled into arsenals that for the most part can be considered surplus stockpiles. It is a case of the virtually useless being regarded as utterly vital. The arsenals are made up of spurious weapons, yet military strategy treats them "as if" they could

25. See Quester (1970). I hedge here and term the arsenals only "essentially useless" because of the deterrence argument. Specifically, some have argued that in the absence of the bomb the United States and the Soviet Union might have engaged in a conventional war, particularly in Europe. Such a war was not *inconceivable,* and the thesis that the bomb deterred it cannot be disproved (or proved!). But one can assert that the nuclear threat is too dangerous to be justified on these grounds alone. When I claim that use of the bomb is not a real political option, I qualify it with the assertion "for the most part," since President Eisenhower did make some nuclear threats. Clearly, the position taken here is subject to controversy: thus American deployments and war plans do envision such things as limited nuclear war at both battlefield and strategic levels. I argue that such plans *rarely* constituted real options. That is, the idea of nuclear paralysis was available from the start and, as a result of nuclear learning, became more ingrained over time. The Cuban missile crisis rendered talk of fighting a nuclear war impermissible in the public culture. President Reagan, to his dismay, violated this taboo and unleashed a war-fighting panic by his trigger-happy talk in the early 1980s.

be used and has imparted a furious impetus to their morbid growth. Indeed, the effort is constantly made to evade the paralysis pole of the paradox, for the idea of omnipotence implies that it is somehow possible to wield such weapons.

On a theoretical plane, the elements of the paradox were grasped in the immediate aftermath of Hiroshima. When Bernard Brodie, the first theorist to address nuclear strategy, warned that the role of the military would now be to prevent wars rather than to win them, he grasped the idea of paralysis. (He later shifted his views, an indication of the radical changes so common in the nuclear trap.)[26] At the same time, scientists warned of an *unlimited* arms race, thereby noting the indispensable aspect of the paradox. Of course, both of these theoretical positions had to be given practical significance, a result of nuclear learning that came from actual experience with the bomb. Whereas Truman never stopped wavering in his attitude toward the bomb, President Eisenhower did try to use American nuclear supremacy diplomatically and coercively in Korea and in the Quemoy and Matsu episode. The concluding remarks of a study of atomic diplomacy during the Korean War bear repeating, since they so admirably assert the need to learn the paralysis inherent in the paradox of indispensability:

> Nuclear weapons were not easily usable tools of statecraft that produced predictable results. One could not move from deterrence through compellence through their possession. . . . The Korean War might be seen as an experience that schooled American statesmen in that practice. It offered not the determinative, but the first, of a series of lessons that would eventually produce full understanding of the paradox of nuclear weapons: They confer upon those who possess them more responsibility for restraint than disposable power.[27]

I will suggest that the definitive lesson in restraint was the Cuban missile crisis, although, paradoxically, it helped fuel the arms race. And even following the missile crisis, the United States would continue seeking means of using nuclear weapons.

In developing the story of the arms race, I seek to unfold the dimensions of this paradox of indispensability. In other words, how

26. On shifts by Brodie and the more general problem of inconsistencies among nuclear strategists, see Herken (1985); Kaplan (1983); and Freedman (1981).
27. Dingman (1988–89, 91).

did "weapons" whose essential prostration became more and more obvious continue to be regarded as indispensable? I suggest that for the most part the sense of indispensability was a panic-induced phenomenon. Nuclearism is a complex and shifting faith, and following each panic there is an attempt to restore the American sense of nuclear omnipotence—to find ways of wielding the power and, ideally, to find "the definitive technological solution that will free us of our fear at last and make us safe and secure in a nuclear world."[28]

The first part of the story surveys nuclear preparations and attitudes during the period of the American monopoly. A counterfactual question that is often posed asks why the Americans did not use the bomb during their monopoly. As we saw in Chapter 4, calls for preventive war at the time were fairly common and were not limited to the lunatic fringe. But what most characterizes American behavior during the monopoly is a real lack of preparation and a failure to match strategy with the reality of the arsenal. The nuclear threat was mostly hollow, as befits a numinous and secret "weapon" imbued with such terror. For Truman it was *at best* a weapon of last resort. Others expressed contradictory fears: that weapons of mass destruction would paralyze the democracies or, in contrast, that the public would insist on their use in any conflict. The greatest irony here, perhaps, is that there was more ambivalence about the winning weapon as a weapon than there was about it as a decisive factor in the political, diplomatic, and economic realms. Thus it was only *after* the moral panic associated with the Soviet atomic bomb that the arsenal was vastly expanded and planning and reality brought more in line with one another.

The dilemma that limited the economic clout of the U.S. military was put succinctly by Bertrand Russell: "Average Americans were oppressed by two fears, fear of Communism and fear of the income tax."[29] As a consequence of the fear of taxes, isolationism, and the relative safety of "fortress America," the political and economic power of the military in the United States prior to World War II was highly circumscribed. The MIC that emerged during the war did not remain at its end, even as the start of the Cold War became apparent.

The bomb, as a result of the secrecy of the Manhattan Project, was not included in the initial postwar military plans. Of the existing military services, only the Army Air Corps was well disposed to it in the first place, warning of "sudden death out of a clear sky."[30] Aside

28. Boyer (1985, 106).
29. Cited in Sanders (1983, 86).
30. Cited in Herken (1980, 196).

from traditional military conservatism, the other services feared that the bomb would be seen as a panacea and used to diminish their budgets. All the services, however, soon came to love their winning weapon.

Reliance on the bomb had less to do with strategic planning than with pragmatics. At the end of the war, the American army melted away. Public sentiment and campaigns for "bringing the boys home" resulted in a more rapid demobilization than had been planned. An armed force of more than 11 million had been reduced to 1.6 million by late 1946. (The Red Army, in contrast, was only reduced by a few million.) At the same time, Congress rejected universal military service and so restricted defense budgets that even the Manhattan Project and the production of atomic bombs and delivery systems suffered.[31] So too did aircraft manufacturing, which, according to the Finletter Air Policy Commission, found civilian demand insufficient.[32] While a vigorous campaign to increase government expenditures had little effect, the communist coup in Czechoslovakia did bring a short-term budgetary boom to the defense industry. As late as 1961, however, its economic viability remained almost as problematic as in 1947.[33]

Fiscal constraint in military matters was abetted by the American atomic monopoly. In 1946, Walter Lippmann viewed it as

> the perfect fulfillment of all wishful thinking on military matters: here is a war that requires no national effort, no draft, no training, no discipline, but only money and engineering know-how, of which we have plenty. Here is the panacea which enables us to be the greatest military power on earth without investing time, energy, sweat, blood and tears, and—as compared with the cost of a great Army, Navy, and Air Force—not even much money.[34]

So the military was prodded, by a variety of pressure groups, into accepting the bomb as its preeminent weapon.

The numinous ambivalence created by atomic weapons affected the early American nuclear stockpile. By June 1947, it amounted to

31. See Herken (1980).
32. Krell (1981).
33. McDougall (1985, 321).
34. Cited in Freedman (1981, 48).

thirteen implosion bombs, each taking thirty-nine people two days to assemble. The lack of preparation was partly due to technical problems, particularly the scarcity of fissionable materials, created by peacetime constraints of "efficiency and economy" imposed on the Manhattan Project. But overshadowing this was a serious lack of policy direction, as well as monopolistic complacency, secrecy, and interservice rivalries. An obsessive secrecy surrounded the stockpile, and the policy elite were not interested, as if they wanted to avoid the "burden of knowledge."[35] The taboo was clear in Truman. In a cabinet meeting of October 1946, he admitted that the United States may have had less than a half-dozen bombs, but quickly added, "that was enough to win a war."[36] David Lilienthal, after assuming the chairmanship of the Atomic Energy Commission (AEC) in the spring of 1947, reported that the stockpile was not "adequate to meet the security requirements of the U.S."[37]

Planning was not tuned to reality. A Joint Chiefs of Staff report, "Strategic Vulnerability of Russia to a Limited Air Attack," two months after the war, selected twenty cities for nuclear attack. It gave no consideration to the number of bombs available or to the problems of delivering them. Significantly, the military pushed for a larger stockpile because production came under the budget of the AEC and thus did not require cutbacks in other military programs. As the military came to accept the idea of deterrence in the summer of 1947, it aimed to have four hundred bombs by the early 1950s.[38]

If military acceptance was retarded, public belief in the revolutionary nature of this weapon, coupled with fiscal constraints, forced the joint chiefs to reevaluate and come to realize its potential. The idea of an enduring supremacy was intoxicating. General Groves held nothing back when he declared that "we are now in a favorable position. . . . We should get our bases now and plan not for 10 years but for 50–100 years ahead."[39] At the same time, planners recognized the danger of paralyzing themselves with reliance on nuclear weapons:

A situation dangerous to our security could result from impressing on our own democratic peoples the horrors of future

35. See Rosenberg (1982).
36. Cited in Herken (1980, 197).
37. Cited in Herken (1980, 197).
38. Rosenberg (1979).
39. Cited in Herken (1980, 112).

wars of mass destruction while the populations of the "police" states remain unaware of the terrible implications.[40]

In early 1948 America's military planing was put to the test. In March, the coalition government in Czechoslovakia was ousted in favor of a Russian puppet government. There was little the United States could do. At the time, the atomic arsenal was still mostly a promise. As Lilienthal later remarked of 1948: "It was assumed that we had a stockpile. We not only didn't have a pile; we didn't have a stock."[41]

April saw the beginning of the Berlin crisis. By 24 June, Russia had blockaded ground access to the Western sector of the city. The United States responded with an airlift and the decision to send atomic bombers to England. However, Truman delayed sending the aircraft until mid-July in order to avoid provoking the Russians. Even then, the bombers were sent, but not the bombs. (Moreover, the planes had not been modified to carry the new bombs, three of which were tested in Operation Sandstone.)

Secretary of the Army Kenneth Royall commented to Truman in the summer of 1948: "We have been spending 98 percent of all the money for atomic energy for weapons. . . . Now if we aren't going to use them, that doesn't make any sense."[42] The president reiterated that the bomb was not a military weapon and refused to hand custody of it over to the military. He added: "You have got to understand that I have got to think about the effect of such a thing on international relations. This is no time to be juggling an atom bomb around."[43] The future did not create any other timely moments for juggling it, either.

The atomic ruse was most likely seen through by the Soviets. Yet Americans were to attribute the eventual "victory" in Berlin to the bomb. Baruch wrote to John Foster Dulles: "The only thing that stands in the way of the overrunning of Europe today is the atomic bomb. . . . When we outlaw that there is nothing to stop the Russian advance."[44] Secretary of Defense James Forrestal wrote:

He [George C. Marshall] never recorded a dissent from the "unanimous agreement" of this dinner meeting. Marshall was

40. Cited in Herken (1980, 221).
41. Cited in Herken (1980, 239).
42. Cited in Herken (1980, 244).
43. Cited in Herken (1980, 260).
44. Cited in Herken (1980, 268–69).

to quote to [Truman] a remark of John Foster Dulles that "the American people would execute you if you did not use the bomb in the event of war"; [Lucius D.] Clay said that he "would not hesitate to use the atomic bomb and would hit Moscow and Leningrad first"; Winston Churchill, going even further, told [the president] that the United States erred in minimizing the destructive power of the weapon—to do so was to lend dangerous encouragement to the Russians.[45]

A State Department memorandum suggested that domestic pressures in wartime "might force the use of atomic weapons, even if the chief executive were inclined against it."[46]

In September 1948, NSC-30 was completed, with its "Policy on Atomic Warfare." It shied away from any public statement on the use of the bomb in warfare, fearing that a debate might convince the Soviets that they could act with impunity. What it did stress was that they "should in fact never be given the slightest reason to believe the U.S. would even consider not to use atomic weapons against them if necessary."[47]

If fiscal restraint and lack of preparation so clearly characterize American nuclear behavior during the period of the monopoly, the obvious question is how the MIC reasserted itself. Here I argue that periods of fiscal restraint in military matters are punctuated by panics produced by spectacular Soviet threats to American nuclear hegemony. In other words, the panics operate as "switchmen" between fiscal constraint and nuclear buildups, and between a sense of paralysis and a sense of indispensability.

The Outbreak of Panic

The first postwar moral panic was unleashed by three (seemingly concatenating) events: the Soviet atomic explosion announced in September 1949, the communist takeover of China one month later, and the Korean War in June 1950. The Soviet atomic bomb came years before officially expected and, despite Truman's efforts to

45. Cited in Freedman (1981, 53).
46. Cited in Herken (1980, 262).
47. Cited in Herken (1980, 268).

downplay its significance, initiated the panic.[48] The collapse of the Nationalist forces in China disillusioned most Americans and, like the Soviet bomb, was blamed on a lack of security.[49] The subsequent invasion of South Korea by the North focused the growing sense of panic, giving credence to the view that, armed with atomic bombs, the Soviets now represented a formidable military (as opposed to political) threat, both in Asia and in Europe.[50]

This conjunction of communist gains clearly created a moral panic.[51] That is, there was a perception of an extraordinary threat to societal values and interests that was presented in a stereotypical fashion by a policy elite with high credibility and perceived morality. These events stoked the Great Fear and led to various diagnoses, solutions, and ways of coping. The panic-induced distortion of threat began with National Security Council Memorandum 68 and culminated in a frantic McCarthyist hunt for traitors. Behind the public display of threat and morality, the resolution of competing interests resulted in a massive increase in the military budget and the atomic stockpile, as well as the H-bomb decision. Following the election of Eisenhower, the panic ended.

The fear generated by the Soviet atomic explosion was palpable, and affinities between nuclear supremacy and both national security and the sanctity of the American way of life were quickly forged. There was a wave of public and private debate over policy toward the Soviets. The drafters of NSC-68 asserted in April 1950 (before the Korean War) that the American people stood "in their deepest peril."

48. Too many important events are compacted into too short a time to term this a pure nuclear panic "caused" by the Soviet bomb. Even a microscopic view of events is unlikely to sort it out satisfactorily. But the direct nuclear component of the panic, seen in the expansion of the arsenal, the H-bomb decision, NSC-68, and the atomic-spy scandals, is certainly significant. The bomb also fed into other events, including support for the Korean War and McCarthyism. In his comments on the Soviet nuclear advance, Truman spoke of an explosion, not a bomb, and 1954 was given as the year of "maximum danger" (the Soviets did not yet have a delivery system, and there was the assurance of U.S. bases circling the USSR). The fear of an enemy bomb had also been "worked out" at the height of the Great Fear (1945-1947). In pure nuclear panics, the nuclear threat is more immediate, direct, and startling.

49. China experts tend to see the communist victory, which undermined Acheson's usefulness as secretary of state and became a major concern in the McCarthyist attack on the State Department, as more important than I do. While the matter cannot be settled here, the China issue has proved to be less enduring than the Soviet nuclear threat, Korea, or McCarthyism in the historical mythology that is part of the common stock of social knowledge. More broadly, the moral panics discussed here seem to have endured over the years: they are still part of the collective mythology of the nation.

50. See Donovan (1982).

51. See Ben-Yehuda (1985).

The memorandum termed communism "a new fanatical faith" and called for its military containment across the globe.[52] Others went further and recommended a preventive war before the Soviets could attain "nation killing capabilities." David Lilienthal recorded the views of Senator McMahon: "What he is talking about is the inevitability of war with the Russians, and what he says adds up to one thing: blow them off the face of the earth, quick, before they do the same to us— and we haven't much time."[53] The Soviet explosion reactivated the Great Fear in part because of the widespread belief that the Soviets, who had massacred so many of their own people, would not let the risk of even millions of Soviet lives deter them from starting a nuclear war.[54]

Where the first announcement of the Soviet atomic bomb had created shocked disbelief, it now appeared that they had somehow managed to get ahead! The Military Liaison Committee wrote that the Soviets' "stockpile and current production capacity are equal to or actually superior to our own, both as to yields and numbers." It further expressed the concern that a Soviet H-bomb "may be in actual production."[55] The Pentagon argued that they were now capable of "producing an effective quantity of bombs in a relatively short period of time." In the event of a nuclear war, the Soviets would blunt the attack by moving industry to a Western Europe occupied by the Red Army: "This course would enable Russia to achieve dominance over all the Eurasian land mass, and place her in position for an early conquest of Africa, and the isolation of the Western Hemisphere for later attack, before U.S. industry could recover."[56] One begins to wonder if even the devils who so wantonly tempted the monasteries of the Middle Ages had magical power anywhere near this great.

Were the elite, as Jerry Sanders suggests, cynically "overselling the threat?" Evidence pertaining to genuine fear and cynical manipulation among the elite remains difficult to come by.[57] Policymakers had started formulating plans for the militarization of national security two years before the panic. They were aware of the need to "sell" their solution to the Soviet threat and regarded the Korean War as a

---

52. See Sanders (1983).
53. Cited in Clark (1980, 258–59).
54. See Freedman (1981, 142).
55. Cited in Herken (1980, 326).
56. Cited in Herken (1980, 313).
57. See Yergin (1977, 209). NSC-68 was a classified document, which renders problematic any claim that it was intended or used to cynically manipulate fear among the public or significant sections of the elite.

particular opportunity to overcome the perceived congressional opposition to increased funding called for in NSC-68. However, the elite's awareness of the need to sell the threat in order to gain political leverage for a particular solution does not mean that its members were immune to the fear that Sanders admits gripped the public. Indeed, after the outbreak of the Korean War, many military and political leaders feared that the Soviets had achieved a situation of strength and were prepared to initiate a general war.[58]

In this regard, there is poll data indicating considerable public support for increased military spending prior to the Korean War.[59] This is in contrast to elite perceptions, which held that communist aggression in Korea was necessary to mobilize public support for budgetary increases called for in NSC-68. Effectively, then, military leaders and the Congress aimed to exploit the fortuitous events that fed the rising panic.

The degree of public involvement in the panic can be inferred from the extensiveness of the spy scandals and what Truman termed the "smear age." In discussing the hysteria, Clarfield and Wiecek assert that the bomb and the witch-hunts were "more than tangentially related."[60] They may in fact be understating the case. As previously observed, the postwar fear of communism was given its initial impetus by atomic-spy stories. The awesome power of the bomb, coupled with the belief in "atomic secrets," fostered a mania for secrecy and security. In 1947, following the Republican congressional victories, which played on the fear of communism, Truman established a Federal Loyalty and Security Program to vet federal employees. By 1953, when the new Eisenhower administration put a more stringent loyalty program in place, more than 4,750,000 persons had been scrutinized by "name checks." Of these, 26,236 were referred for examination by loyalty boards; 560 of them were ultimately dismissed or denied jobs as a consequence of "adverse findings."[61]

The roving hysteria turned to Alger Hiss, who was accused of passing on government secrets. His two trials for perjury were a national sensation. Much of the liberal establishment came to his

58. See Brodie (1973, 57–65) and Trachtenberg (1988–89).
59. See Huntington (1962). However, data on public attitudes are quite limited. Public-opinion polls reveal what McClelland (1977) terms the "Cold War pattern." A generalized sense of fear was sustained between 1948 and 1964, but the polls are conducted too infrequently and are not sufficiently sensitive to capture sudden and extreme surges of fear.
60. Clarfield and Wiecek (1984, 143).
61. See Phillips (1975, 368).

defense: Truman termed the Hiss hearing a "red herring," and Secretary of State Acheson remarked that "I will not turn my back on Alger Hiss."[62] After Hiss was found guilty (and his guilt or innocence remains the subject of controversy), Acheson lost much of his effectiveness. In the *New York Times Index* for 1949, the section entitled "U.S.—Espionage, Sabotage, Treason, and Subversive Activities" fills seventeen pages of very small print.[63]

This brewing hysteria took on its most virulent form in the context of the first postwar moral panic. Joseph R. McCarthy's tirade against the Truman administration dates from February 1950, when the senator sent a telegram in which he claimed to have the names of 57 communists in the State Department. McCarthy demanded that Truman hand over to Congress the files of persons at State listed as poor risks by the Loyalty Security Board. This was a direct challenge to the president, who had ordered that the files be kept secret to avoid unwarranted intrusions by the House Committee on Un-American Activities and other subversive-hunting bodies. So began the rounds of accusations and denials, with McCarthy pulling dossiers but not putting names to the charges.

Truman sought to ride out the onslaught. Of these events, he later wrote, "We are just going through one of those hysterical stages and we may be better off when we come out of it."[64] He went so far as to have his staff prepare a report on these hysterical episodes, beginning with the Alien and Sedition Acts and extending through the Anti-Masonic Movement, the Ku Klux Klan, and the Red Scare. But Truman underestimated the threat. (In a memorandum, the president termed McCarthy a "ballyhoo" and advised, "I don't think you need pay any particular attention to him.")[65] McCarthy's demagogic onslaught brought him national and international attention. So vehement was his campaign that Truman soon agreed to provide a congressional subcommittee the Security Board files on eighty-one persons accused by McCarthy. Contrary to expectations, the hysteria did not burn itself out.

Accusations, typically with little or no substance, were made against prominent political leaders, bureaucrats, academics, and artists. To put it another way, the hysteria was not directed, as was so often the case, against working-class movements, foreigners, or rebellious Afro-Americans. Yet the poisoned atmosphere was not limited to the

62. Cited in Phillips (1975, 373).
63. See Boyer (1985, 103).
64. Cited in Donovan (1982, 166).
65. Cited in Donovan (1982, 166).

nationally prominent. A number of state governments established their own "little HUACs,"[66] and many communities set up inquisitions to purge school texts and public libraries of "Un-American doctrines." Blacklists, loyalty oaths, and harassment became the order of the day. Even the army was attacked; and it was here that McCarthy overstepped himself and lost sufficient credibility that the Senate eventually censured him.

McCarthy's telegram to Truman was sent during the same month that the British arrested Klaus Fuchs, a German scientist who had worked at Los Alamos, for passing on atomic secrets to the Soviets. This correspondence is probably coincidental; but in its aftermath, the two events converge. The arrest of Fuchs gave the imprint of truth to the many spy charges and conspiracy theories that were in the air. The ensuing hysteria removed the restraints that usually keep political affairs in tolerable bounds. As early as 1948, HUAC began to investigate and harass scientists with Communist associations. Now the search for "the Alger Hiss of science" was set in motion.[67] McCarthy's oft-repeated charge that lax security procedures had permitted the Soviets to "steal" the atomic bomb gained wide currency.[68] The big-game hunt led to the arrest, conviction, and execution of Julius and Ethel Rosenberg for passing on atomic secrets to the Soviets. The hunters also pursued larger quarry. They took aim at Robert Oppenheimer, removing his security clearance as a result of his youthful Communist associations and his opposition to the H-bomb. With the overlapping threats, the United States created by 1950 a formidable security apparatus, "staffed with civilian and military secret police, fed by professional informers, supplemented with concentration camps (actually never used) waiting to receive political radicals. . . ."[69]

In sum, the timing of the panic was a result of three communist advances that created a surge of fear through the public and possibly much of the elite. The latter regarded their efforts to sell the threat as contingent on actual events, and aimed to sell it not so much to create fear as to direct that fear toward particular solutions. Among the elite, the affinity between nuclear superiority and national security had been forged before the panic and was soon rendered unassailable (despite efforts by George Kennan and others). Now the policy elite

66. See Phillips (1975).
67. See Clarfield and Wiecek (1984, 144–46).
68. See Powaski (1987, 57–58).
69. Clarfield and Wiecek (1984, 143).

and the MIC publicly directed the fear along the lines of containment militarism, which provided the legitimation for Truman's reversal of his fiscal conservatism and the inauguration of a massive nuclear buildup.

So great was the president's ambivalence about the bomb that he did not make it the centerpiece of American strategy until the spring of 1949.[70] But it was only after the announcement of the Soviet explosion, which was accompanied by scientific, congressional, and public pressure for the nation to quickly reassert nuclear superiority, that Truman agreed to build reactors capable of supporting an enormous expansion of the stockpile.[71] Where strategic plans before the Soviet explosion had called for the availability of 400 atomic bombs by the early 1950s, the stockpile grew to about 1,000 weapons in the summer of 1953, to nearly 18,000 by the end of the decade, and to about 30,000 by the mid-1960s.

The loss of the monopoly undermined both American military strategy and the sense of the country's uniqueness embodied in the idea of a sacred trust. Where the atomic monopoly had been proof positive of America's uniqueness, a degree of special standing was preserved by the belief that the Soviets had stolen the secret of the bomb. Moreover, the idea of a monopoly was transformed, "through sleight of hand,"[72] into the idea of nuclear supremacy. Supremacy was still enough to make America seem exceptional, especially in the race for the H-bomb (the "super" bomb). The atom bomb had not proved to be omnipotent in the diplomatic, military, or political realms; it had inspired as much fear as confidence. But with the loss of the secret, the mantle of omnipotence was resurrected and transferred to the superbomb. Here was a thermonuclear weapon with no theoretical limit on its destructive power. Staying ahead—proving and retaining one's omnipotence over the devil—was a potent motivation and impetus.

The film was a rerun, the same attempt to gain what could not amount to more than a temporary lead. In this case, the drive for a transitory advantage created a perpetual menace to humankind and to the very order of nature itself. Nuclear omnipotence would now be associated with exterminism. The idea of omnipotence as exterminism is well formulated by Gunther Anders:

70. Rosenberg (1979, 75).
71. See Pringle and Spigelman (1981, 100); Rosenberg (1979, 79; 1982; 1983, 23).
72. The phrasing is used by Wolfe (1984, 94).

> If there is anything that modern man regards as infinite, it is
> no longer God; nor is it nature, let alone morality or culture;
> it is his own power. *Creatio ex nihilo,* which was once the mark
> of omnipotence, has been supplanted by its opposite, *potestas
> annihilationis* or *reductio ad nihil;* and this power to destroy, to
> reduce to nothingness, lies in our hands. . . . It is we who are
> the infinite.[73]

While the idea of exterminism is embedded in the H-bomb, its full
realization would require the hair-trigger technology of ballistic missiles. And even as the practical conditions for exterminism were
realized, there was still the attempt to wield these weapons, to render
them usable in limited ways so that deterrence could be extended to
Europe and to smaller, brushfire wars.

In many respects, the story of the H-bomb parallels that of the
A-bomb. Like its predecessor, the H-bomb was shrouded in secrecy.
As with the atom bomb, the idea of a thermonuclear weapon held a
grim attraction and fascination for some scientists. Edward Teller, the
prime scientific mover behind the H-bomb, claimed that several
scientists were recruited to Los Alamos "only because they were
intrigued by the thermonuclear possibilities."[74] The "genesis complex," so visible at Trinity, also extended to the fusion bomb. When
Oppenheimer showed Ernest Lawrence a vial of clear liquid that
looked like water, the latter wrote:

> It was the first highly diluted minute sample of superheavy
> water, composed of tritium and oxygen, ever to exist in the
> world, or anywhere in the universe, for that matter. We both
> looked at it in silent, rapt admiration. Though we did not
> speak, each of us knew what the other was thinking. Here was
> something, our thoughts ran, that existed on earth in gaseous
> form some two billion years ago, long before there were any
> waters or any forms of life. Here was something to return the
> earth to its lifeless state of two billion years ago.[75]

As with the atom bomb, the politicians were advised to proceed
with caution, but again they ignored their advisers. Detailed consideration of the H-bomb and other possible responses to the Russian

73. Cited in Chernus (1986, 31).
74. Cited in Clark (1980, 261).
75. Cited in Clark (1980, 262).

bomb was undertaken by the General Advisory Committee (GAC) of the AEC. The scientists who comprised the GAC unanimously advised in late October against an all-out effort to produce the superbomb. Their logic, as Herbert York details it, was compelling.[76]

On the one hand, there were the problems of the super itself. Not only was it merely a theoretical possibility; there was no certainty that it could be turned into a deliverable weapon. And even if it could, there was the issue of what to do with it. The hydrogen bomb could not realize the aims of the belligerents. Too big to be used against any possible military target, it amounted to a "weapon of genocide" against civilian populations. The report read: "We base our recommendations on our belief that the extreme dangers to mankind inherent in the proposal wholly outweigh any military advantage that could come from this development."[77]

On the other hand, the H-bomb was not required to assure U.S. security. A number of atom bombs could serve the same purpose, especially against the Soviet Union with its dispersed population. The committee wrote: "Should they use the weapon against us, reprisals by our large stock of atomic bombs would be comparatively effective to the use of a super."[78] To further assure security, the committee did recommend expanded production of fissionable materials, the development of smaller tactical weapons, and continuation of the booster program, whereby atom bombs were made far more destructive. At the time, the United States was near to producing a fission weapon with a yield equivalent to 500,000 tons of TNT, a large jump over the 20,000-ton bomb used on Hiroshima.

The committee saw "a unique opportunity of providing by example some limitations on the totality of war and thus of eliminating the fear and arousing the hope of mankind."[79] Implicit in this is the notion that some form of international control might be more feasible now that Russia had the bomb. No longer need that state bargain from a position of inferiority—parity provided a possible basis for agreement.

While Lilienthal backed the GAC position, he was appalled by the growing strength of the pro-super forces. Commenting on a report about scientists at Berkeley and Los Angeles, he noted: "[there] is a group of scientists who can only be described as drooling with the prospect and bloodthirsty."[80] He worried that "just as the A-bomb

76. See York (1970, 1976).
77. Cited in Clark (1980, 265).
78. Cited in Clark (1980, 266).
79. Cited in Kevles (1978, 378).
80. Cited in Pringle and Spigelman (1981, 97).

obscured our view and gave a false sense of security, we are in danger of falling into the same error again in discussion of [the H-bomb]—some quick and easy way out."

The GAC debate had been secret. The idea of the bomb was not. Where some churches, scholars, and editors sounded warnings against the new danger, most of the pressure was in favor of the bomb: "The public clamor for the 'super' became a roar."[81] A State Department survey of public opinion in March stated that most respondents supported "the adoption of stronger U.S. foreign policy measures." Without such measures, there was the real possibility of "increasing public pressures which could become dangerous for some sort of bold action."[82] The public wanted a clear response to the Soviet atomic breakthrough. So too did congressional and military leaders. Truman worried that if he did not proceed with the project, Congress would appropriate funds on its own to build the super.

If McCarthyism and the trial of the Rosenbergs are the most vivid and easily recalled aspects of the advancing moral panic, many politicians and military leaders helped fuel it by engaging in apocalyptic prophecies. Senator Brien McMahon sounded the alarm:

> If we let Russians get the super first, catastrophe becomes all but certain—whereas, if we get it first, there exists a chance of saving ourselves. . . . Total power in the hands of total evil will equal destruction. . . . In my judgment, a failure to press ahead with the hydrogen bomb might well mean unconditional surrender in advance by the U.S. to alien forces of evil.[83]

All this, despite the senator's concession that

> There are scarcely more than two or three urban targets in all Russia which measure up to the destructive power of [the superbomb], and they could be thoroughly attacked with ordinary atomic bombs.[84]

The Joint Chiefs of Staff originally argued that the bomb was not a weapon and did not press for it.[85] The super was not needed for most

---

81. Pringle and Spigelman (1981, 100).

82. Cited in Herken (1980, 327).

83. Cited in York (1970, 60).

84. Cited in Herken (1980, 316).

85. See Rosenberg (1979) for more detail on the shifting military response to the H-bomb.

targets in the USSR, and it did not meet the limited war objectives of the United States. Omar Bradley, chairman of the joint chiefs, acknowledged that the super would not confer any military advantage. However: "Possession of a thermonuclear weapon by the USSR without such possession by the United States would be intolerable" both by its "profoundly demoralizing effect upon the American people" and by the "tremendous psychological boost" it would give to Soviet leaders.[86] In the same vein, Paul Nitze, head of the Planning Division of the State Department, held that it was absolutely necessary for the world to continue believing in the superiority of American technology.[87] Here at least there is some correspondence between the reality and the conclusions drawn.

Truman decided to proceed with the H-bomb in January 1950. But as a result of mounting pressures, he ordered a crash program in March of that year. In November 1952, the United States exploded the first thermonuclear device, with a yield of some 10 million tons of TNT. Two weeks later the United States tested the booster A-bomb, with a yield close to the predicted 500,000 tons of TNT. Russia exploded a small thermonuclear device in August 1953, with a yield that may have been less than that of the booster A-bomb. In March 1954, the United States tested six H-bombs over a ten-week period. In November 1955, Russia tested its first superbomb. In summarizing the Soviet effort, David Holloway observes that it displays both an internal dynamic and a response to American actions.[88]

The Korean War was a surprise. The country was not of vital interest to the United States, and in a speech on 12 January 1950 Secretary of State Acheson did not include Korea in the "defense perimeter" against communism.[89] The decision to commit American troops certainly owed a great deal to the Soviet atomic bomb, particularly the fear that the Soviets would be emboldened by it.[90] The perceived links between the Soviet bomb and the Korean War are apparent in the trial of the Rosenbergs. Sentencing them, Judge Kaufman said that their having given the Soviets the atomic bomb, years before pre-

---

86. Cited in Herken (1980, 316–17).
87. See Jungk (1958, 255–56).
88. See Holloway (1983, 23–27).
89. Fleming (1961, 592–93).
90. Donovan (1982) observes that the reversal of U.S. policy was based on the belief that Korea was the prelude to further communist aggression. In the first week of December 1950, the Joint Chiefs of Staff were convinced that a Soviet initiation of total war in Europe was only weeks away (Brodie 1973, 57–65).

dicted, caused the communist aggression in Korea, with resultant casualties exceeding fifty thousand.[91] President Eisenhower used the presumed link between the Rosenbergs' espionage and the Korean War to deny them clemency.

Before the panic, the fiscally conservative Truman kept a ceiling of around $14 billion on military budgets, even as his warring services made requests totalling $30 billion. Anything more than $15 billion was thought to be "grievously inflationary."[92] Indeed, the president repeatedly stated that he hoped to cut military spending down to $5 to $7 billion.[93] Following the outbreak of the Korean War, Truman increased the military budget into the $50 billion range. But where the H-bomb and Korean War decisions relied on elite consensus, much of the increased funding called for in NSC-68 evoked elite conflict. Specifically, the memorandum stressed the gap between the goal of containment and the military weakness of the United States.[94] Large parts of the increased appropriations in 1950 were not for the Korean War but for the general expansion of U.S. military power.[95] Especially controversial were plans to bankroll Europe's rearmament, garrison American troops there, and restore universal military service. Consensus on these issues was difficult to achieve in the administration (for example, Secretary of Defense Johnson, a fiscal conservative, objected to them), and the plans were opposed by conservative Republicans and business elites who were against tax increases and wanted to limit the government's power to regulate the economy.[96]

The ensuing debate involved efforts by competing elites to give direction to the surging panic provoked by the concatenating communist successes. Ultimately, Congress endorsed increased spending and the garrisoning of troops in Europe, but not universal military service. Here the contrast between the nuclear and conventional expansions is instructive. Whereas spectacular Soviet nuclear threats engender a semiautomatic consensus in favor of nuclear buildups,

91. Hyde (1980, 186).

92. Donovan (1982, 59).

93. Krell (1981, 230).

94. Truman's budget ceilings resulted in all services being below the manpower strength authorized by Congress, as well as in obsolete equipment, inadequate supplies, and short training periods (Donovan 1982, 254–55). I focus on the nuclear realm, and since this has required 15 to 25 percent of military outlays, the overall military budget is not a major concern. Problems in interpreting military expenditures are discussed by Abolfathi (1980) and Mintz and Hicks (1984).

95. Krell (1981) and Trachtenberg (1988–89).

96. Lo (1982).

conventional threats typically generate greater elite and public debate. In this realm, where the affinities associated with nuclear supremacy do not exist, the United States has readily accepted military inferiority, as illustrated by the persistent and much-touted conventional gap in Europe.

6

Sputnik and the Challenge to America's Destiny

The panic ended as the Korean War became a holding action and General Dwight Eisenhower was elected president. But the panic had massive consequences. The military posture and preparedness of the United States were irreversibly transformed, particularly in the nuclear realm. Where Truman started as a fiscal conservative and overrode military objections to major defense cuts in the immediate postwar period, his final defense plan called for a doubling of expenditures to $80 billion annually. The new Eisenhower administration, however, curtailed the effects of the panic and put a ceiling of $40 to $45 billion on defense.[1] In many respects, the time seemed right for toeing the line. The relationship between the American and Soviet camps was stalemated. The Korean War was a standoff, and public support for it waned as it become clear that victory was impossible. No progress was being made in settling the outstanding issues in

1. Brune (1985, 22).

Europe; an indefinite standoff seemed likely here as well. The new president was immensely popular and had a signal advantage over his predecessor: as a five-star general who had been the Supreme Commander of Allied Forces in the European theater during the Second World War, Eisenhower commanded respect for his knowledge and was not easily cowed by military leaders claiming that national security would be jeopardized if their pet projects were not funded.

Eisenhower, perhaps even more than Truman, was a fiscal conservative. The spending spree envisioned in NSC-68 ignored what he saw as the vital "connection between national security and fiscal responsibility."[2] Inflation, not the Soviet Union, was the most insidious danger. His goal was to reduce the military budget without compromising the ability to contain communist aggression. To this end, Eisenhower sought to rationalize defense planning and budgets, both of which ran amok in the last years of Truman's tenure. His strategy was to plan for "the long haul," rather than to constantly react to immediate crises. He reorganized the Joint Chiefs of Staff and instructed them to formulate a military strategy that would rely on the nuclear arsenal to deter Soviet aggression.

Following the Paris designers who lowered hemlines on women's clothes, the press dubbed Eisenhower's strategy "The New Look." It promised more for less, or "more bang for the buck." The nuclear arsenal, hitherto unusable, was now to become the weapon of first resort. The policy, as propounded by Secretary of State John Foster Dulles, held that the U.S. response to aggression would "depend primarily upon a great capacity to retaliate, instantly, by means and at places of our own choosing."[3]

Eisenhower, in effect, was seeking to evade the paralysis pole of the paradox of indispensability and to render nuclear weapons usable. Exactly what "massive retaliation" meant, however, was the subject of considerable controversy. Did it really imply massive preemption? Would any brushfire war, as in Korea, call for instant nuclear retaliation? Some journalists and politicians (including Richard Nixon) thought so. Dulles, however, backed away from this extreme position. He suggested instead that the provocations which might induce a nuclear response best be kept secret. The Soviets should be kept guessing. But Basil Lidell Hart, the British military strategist, questioned whether "any responsible government" would "dare to *use* the H-bomb as an answer to local and limited aggression."[4] Massive

2. Cited in Brown (1968, 66).
3. Cited in Clarfield and Wiecek (1984, 155).
4. Cited in Clarfield and Wiecek (1984, 156).

retaliation, he claimed, lacked credibility: it was too inflexible; it was morally irresponsible; it mistook a weapon of last resort for one of first resort; and it was unrealistic, mistaking a short-term advantage for a "long-haul" policy. The Soviets were striving to build their own nuclear deterrent. Once in place, the ideas of massive retaliation would become a threat of mutual suicide.

Yet massive retaliation became official American and NATO strategic policy. To offset a 25 percent reduction in military personnel, "tactical nuclear weapons" were provided to field units in Europe. This reflected a revolution in nuclear technology: bombs were made small and light enough to be used by fighter aircraft, and "techniques and procedures" were developed for their use on the battlefield.[5] Eisenhower, more than Truman, sought ways of employing the weapons. Although he stated repeatedly that a main goal of his administration was to reduce the threat of nuclear war, he also tried to imbue nuclear weapons with an aura of conventionality: "When these things are used on strictly military targets and strictly for military purpose, I see no reason why they shouldn't be used just exactly as you would use a bullet or anything else."[6] Eisenhower did try to use the weapons diplomatically and coercively to end the Korean War, and considered their use in Indochina in 1954 and in Matsu and Quemoy in 1955 and again in 1958. Whether the president was bluffing, and whether the doubtful effectiveness of these nuclear threats was deliberately exaggerated in claims made by the administration, remain subjects of controversy.

Under the auspices of massive retaliation, the Eisenhower administration reduced the defense budget by 20 percent between 1953 and 1956. Military commanders chafed at the cutbacks and constraints. Even the air force, which fared best under the new policy, adopted a rancorous posture. Air force policy called for nuclear supremacy. It wanted to develop the capacity to destroy the Soviet bomber force on the ground in a single preemptive strike while maintaining a reserve capacity to attack and destroy Soviet society.[7] But the administration held firm, sticking to the idea of "sufficiency." According to Eisenhower, "there comes a time . . . when the destructiveness of weapons is so great as to be beyond imagination, when enough is certainly plenty, and you do no good, as I see it, by increasing these numbers."[8]

5. See Trachtenberg (1988–89, 29–30).
6. On Eisenhower and arms control, see Powaski (1987, 74–92). Citation in Freedman (1981, 78).
7. See Clarfield and Wiecek (1984, 159).
8. Cited in Clarfield and Wiecek (1984, 159).

The H-bomb, however, inaugurated still another era in the nuclear world. Because it could be made smaller and lighter than the A-bomb yet deliver far more bang, it was well suited for ballistic missiles. That race was starting up. A 1951 memo written by Deputy Secretary of Defense Robert Lovett observed that progress was being made in the development of strategic, submarine-launched, tactical ground-support, and antiaircraft missiles.[9] The Soviets had begun a missile-development program of their own even earlier. Success in these endeavors would deprive America of its most important basis for security: its relative geographic isolation from the Eurasian continent.

Having lost their nuclear monopoly, Americans wanted nuclear superiority. The idea of superiority, however, no longer entailed the notion of a sacred trust: neither the term nor anything equivalent to it appears again in nuclear talk. Rather, technological *competition* in the atomic and missile fields became the focal point of the rivalry between the two camps, especially as the Soviets sought to demonstrate "socialist superiority" by making an all-out effort to prevail in that competition.[10] America's exceptionalism was now equated with the technological superiority associated with liberal democracy and capitalism. In this process, sacred beliefs about the destiny of the nation were no longer limited to ceremonial proclamations. These beliefs were concretely and unequivocally invested in the technology of the arms race, and the two rivals vied in front of the whole world for triumphs that involved not only prestige but the issue of which system would predominate historically.

The "atoms-for-peace campaign" of 1953 represents one effort by Americans to manage nuclear contradictions and to establish technological and historical preeminence. The "fearful effect" of the Soviet atomic bomb had been rendered more fearful by the development of the H-bomb. There was the fear of an "engineering gap" in favor of the Soviets, and Eisenhower was wary of "global humiliation" if Russia "beat us at developing the peaceful side of the atom."[11] He was seeking to foster an image of the United States as more peaceful and more technologically advanced than the Soviets. This image was seen as a key to gaining influence over nonaligned nations, for whom nuclear power was regarded as a solution for basic economic problems. Lewis L. Strauss, chairman of the AEC, captured the thrust of the American vision as follows:

9. Clarfield and Wiecek (1984, 142).
10. See McDougall (1985, 52).
11. Cited in Lowen (1987, 472).

Transmutation of the elements—unlimited power, ability to investigate the working of living cells by tracer atoms, the secret of photosynthesis about to be uncovered—these and a host of other results all in 15 short years [from 1954]. It is not too much to expect that our children will enjoy energy too cheap to meter—will know of great periodic regional famines only as matters of history—will travel effortlessly over the seas and under them and through the air with a minimum of danger and at great speeds—and will experience a lifespan far longer than ours, as disease yields and man comes to understand what causes him to age. This is the forecast for an age of peace.[12]

Effectively, the "image of the future" being incorporated in American civil religion is becoming less transcendental and more nationalistic, drawing on the values of the American way of life. The Good is being equated with technology as such.

Sputnik and the Discrediting of America's Special Destiny

From the Korean truce until *Sputnik,* an element of "normalcy" returned to U.S. society as the economy expanded, the military shrank, and international tensions eased.[13] The death of Stalin in 1953 produced a thaw in the Cold War. From 1954 onward, the Americans and the Soviets engaged in a series of talks on disarmament and a test-ban treaty.[14] Talks on the latter were promoted by the fear of radioactive fallout, stoked in particular by the American Bravo test of 1954, when fallout accidentally contaminated the *Lucky Dragon,* a Japanese trawler. According to Robert Divine, the "'awesome fireball" of the H-bomb created a horror that was too great to face directly. Hence public attention was directed at fallout—the fear of strontium 90—rather than at nuclear proliferation.[15] But even as the peace movement grew in numbers and became internationalized (in

12. Cited in Prins (1982, 44).
13. McDougall (1985, 124).
14. See Divine (1978) and Powaski (1987, 74–92).
15. As a child of about ten years of age, I distinctly recall the fear of strontium 90 and the call to stay indoors on days when radioactive clouds were supposed to be passing over Montreal.

1957 a conference of scientists met in Pugwash, Nova Scotia, to consider the nuclear threat), both the size of the bombs and the numbers tested increased. Public fears were also undoubtedly raised by the publication, in 1957, of Nevil Shute's novel *On the Beach*.

In 1955, Eisenhower proposed an "open skies" plan, which called for aerial inspection of the military installations of each nation. Khrushchev termed the proposal a very transparent espionage device and turned it down. He had good reason to do so, for the Soviets were about to engage in a major bluff. In a June 1955 air show, so many squadrons of Bison intercontinental bombers were seen flying overhead that it appeared as if they had a formidable force capable of striking at the United States. As it turned out, they had about ten of these planes, which repeatedly passed over the reviewing area. The deception worked. Soon congressional Democrats and top air force brass were charging that the administration, in its complacency, had allowed a "bomber gap" to develop. But despite Senate air-power hearings and a parade of doom-predicting generals, a panic did not ensue for several reasons. First, bombers were not a new phenomenon, and their threat potential was not sufficiently spectacular to provoke the Great Fear.[16] Second, the policy elite was not united, with the army and the navy opposing the idea of a gap. Admiral Arthur W. Radford, chairman of the joint chiefs, asserted that "there is good reason to believe that we normally overestimate Communist capabilities in every respect."[17] Eisenhower knew the bomber gap was a bluff and that his officers were employing their own form of bluff to engender support for budgetary increases. The upshot was an extra $1 billion appropriation for the air force and a more rapid procurement of the new B-52 bomber. In 1956, the Soviets had a force of about 150 intercontinental bombers; the United States had some 1,400 B-47 bombers and was beginning to deploy the first of a planned 600 B-52s.[18]

Sputnik proved to be different. In 1955, Nelson Rockefeller, special assistant to the president on government operations, approved a satellite-program report prepared by the National Security Council and added that the achievement of a satellite "will symbolize scientific

16. Panic over bombers was played out in the years before World War II, when they were still a new phenomenon and experts believed that strategic bombing would bring about an immediate collapse of civilian morale and production (see Brodie 1959). The bomber gap also involved a simple quantitative superiority, as opposed to a new, qualitative one, like *Sputnik*.

17. Cited in Prados (1982, 57).

18. See Clarfield and Wiecek (1984, 162).

and technological advancement to peoples everywhere. The stake of prestige that is involved makes this a race that we cannot afford to lose."[19] But President Eisenhower, who civil religionists have correctly characterized as taking a priestly and hence reassuring role, remained a conservative who upheld American impulses against federal control of education and research, unconstrained military spending, and social engineering.[20] Eisenhower was out of touch with the changing realities and would spend his second term fighting a rear-guard action against the technocratic threat to traditional American values unleashed by *Sputnik*.

In August 1957, the Soviet Union claimed that it had conducted the first successful test of an intercontinental ballistic missile (ICBM). The announcement was received with skepticism in the United States. In October, the Soviets launched the first artificial satellite, and less than a month later *Sputnik II* carried a dog into orbit. An American attempt to launch a satellite in December was unsuccessful, with the missile exploding on the launchpad. But even the successful satellite launch in early 1958 did little to diminish the shocking impact of *Sputnik*. The Russians were first into space, and their venture was based on a booster rocket of startling power: one of their satellites orbiting the earth weighed 2,900 pounds.

Almost as much as the atomic bomb, the 4 October launch of *Sputnik* ushered in a new world. Americans rushed outside to view its light soaring through the dark. World attention was riveted on the new phenomenon: "An NBC announcer said, 'Listen now for the sound which forevermore separates the old from the new,' and millions heard the eerie *beep—beep—beep* from outer space."[21] Two transmitters aboard the device were sending continuous radio signals to earth. A new word—*sputnik* is Russian for "traveling companion"—had entered the vocabulary. There was indeed a sense of a pre- and a post-*Sputnik* world.

The Eisenhower administration greeted the Soviet achievement with a studied calm. On 9 October the president explained that the United States could have launched a satellite had it wished. Rear Admiral Rawson Bennett, Chief of Naval Operations, described the satellite as "a hunk of iron almost anybody could launch."[22] But the effort to pooh-pooh *Sputnik* was to no avail. Despite the justifiable claim that the American missile program was on course and that the

19. Cited in McDougall (1985, 120).
20. On civil religionists, see Marty (1974). On spending, see McDougall (1985, 73).
21. Beschloss (1986, 148).
22. Cited in Clarfield and Wiecek (1984, 163).

Soviet spectacular did not mean that they were ahead, the satellite ignited the second postwar moral panic. In this case, it was clearly a panic of omnipotence. Ballistic missiles were a totally new phenomenon. Able to reach the United States in thirty minutes or less, they created an unprecedented peril: the risk of an instantaneous attack, against which there was neither warning nor defense, that threatened the physical annihilation of the nation. Defense Secretary Charles Wilson's joking remark, "Nobody is going to drop anything on you from a satellite while you are asleep, so don't worry about it,"[23] went to the heart of the American terror. Yet this military threat, made concrete in the fear of a bogus "missile gap," was only one element of the *Sputnik* shock. Beyond national security, the Soviet success challenged fundamental assumptions about the technological *and* moral superiority of America, as well as its sense of historical transcendence. Such was the surprise that there was an abortive attempt to explain the Russian lead by their having stolen satellite secrets from the United States.[24]

As the last vestige of security was undermined by the apparent shift of omnipotence to the absolute enemy, confidence plummeted and soul-searching reactions were set in motion. According to James Killian, special secretary for science and technology, *Sputnik* "did violence to a belief so fundamental that it was almost heresy to question it: a belief I shared that the United States was so far advanced in its technological capacity that it had in fact no serious rival." Clare Boothe Luce termed the *beep* of the satellite "an intercontinental outer-space raspberry to a decade of American pretensions that the American way of life was a glittering guarantee of our national superiority."[25]

As the panic picked up the press, Congress, and the public in its whirlwind, the administration stopped downplaying the Soviet advance and shifted to a strategy of reassurance and accelerated American missile programs. Eisenhower, in his memoirs, reveals that the Soviet achievement was "impressive" and that the power of their booster rocket came as a "distinct surprise" and provided them "a spectacular head start." But he was even more startled by the "wave of near-hysteria" and the psychological vulnerability of Americans— "Most surprising of all, however, was the intensity of the public concern." The Soviet feat

23. Cited in Clarfield and Wiecek (1984, 163).
24. See Fleming (1961, 887).
25. Citations in Killian (1976, 3, 8).

precipitated a wave of apprehension throughout the Free World. Newspaper, magazine, radio and television commentators joined the man in the street in expressions of dismay over this proof that the Russians could no longer be regarded as "backward," and had even "beaten" the United States in a spectacular scientific competition. People now recalled with concern that only a few weeks earlier the Soviet Union had claimed the world's first successful test of a multi-stage ICBM— a shot which, the Russians said, demonstrated that they could fire a missile "into any part of the world."[26]

So began what Walter McDougall terms a "media riot," which quickly spread to elements of the elite and the public.[27]

Stereotyped fears and moral recriminations were echoed by others. There were cries of "We have been asleep at the switch"; Senator Henry Jackson demanded a "National Week of Shame and Danger."[28] America was compared to "an overripe, over indulged Babylon." A nearly endless stream of sermons and articles worried about complacency and purposelessness. There were fears that the American economy was stagnant, even as the Soviet Union enjoyed a boom. Joseph and Stewart Alsop described *Sputnik* as the "worst single piece of news of the postwar years."[29] Edward R. Murrow noted that the "acute depression" produced by *Sputnik* and the exhilaration resulting from the launch of *Explorer* were "symptoms not usually associated with normal health, in individuals or nations."[30] To the *New York Times*, America was in a "race for survival." According to Eric Goodman:

> Throughout the United States . . . a sense of alarm, exasperation, humiliation, and confusion mounted. Sputniks I and II dramatized as nothing else could have done that the chief thing on which Americans had depended for their national security and for victory in competitive coexistence with Communism— the supremacy of American technical know-how—had been bluntly challenged.[31]

Where it had been axiomatic that the United States was morally superior, mightier, and blessed with superior institutions, the Soviets

26. All citations from Eisenhower (1965, ch. 8).
27. McDougall (1985, 145).
28. Cited in Beschloss (1986, 148).
29. Clarfield and Wiecek (1984, 164).
30. Cited in Powaski (1987, 68).
31. Cited in Brown (1968, 117).

had won a critical race that enhanced their prestige across the globe. In effect, the *Sputnik* panic challenged the accepted affinity between American nuclear supremacy and national security, on the one hand, and American technological superiority and the country's special historical mission as the leader of the Free World, on the other. As one historian put it, *Sputnik* "registered the end of the 'American Century' that Henry Luce inaugurated in 1941."[32]

Foreboding, mortification, and recrimination were the first reactions to *Sputnik*. Khrushchev's claims of Soviet power took on new depth with the apparent discrediting of America's most fundamental beliefs concerning its future and destiny.[33] Ceding a technological lead to the Soviets meant, according to Edward Teller, "that there is very little doubt who will determine the future of the world."[34] The American humiliation was indeed global in scope. A Japanese newspaper called *Sputnik* "a Pearl Harbor for American Science." In Europe, where the Soviets made a major attempt to cash in on their exploits, confidence in the United States plummeted. Khrushchev used the Soviet feat to engage in what became known as "*Sputnik* diplomacy." He began a campaign of "missile-rattling" that appeared to work in the 1956 Suez crisis. Whereas in 1955 only 6 percent of West Europeans thought the United States weaker than the USSR in military might, in 1960 a plurality of Europeans in every country expected the USSR to be stronger after twenty years of "competition without war."[35] With the new Soviet credibility, not only science and technology but also education, race relations, and economic development became points of comparison between communism and democracy.

The ensuing panic touched on all aspects of American life. The recriminations gave rise to a sense of sacrifice and rebirth. According to Vannevar Bush, *Sputnik* was "one of the finest things Russia ever did for us. . . . It has waked up this country."[36] There was a felt need to reinvigorate American life and to prove the superiority of American institutions. The Rockefeller Panel, one of many goal-setting

32. Cited in Fleming (1961, 885).
33. Both the "missile gap" and the "space gap" were bluffs carried out by Khrushchev on the basis of Soviet secrecy. With the exception of booster size, the American missile program was superior in every respect to the Soviet one (McDougall 1985, 250–62).
34. Cited in Killian (1976, 8).
35. McDougall (1985, 240–41).
36. Cited in McDougall (1985, 153).

study groups, brought together hundreds of "leading citizens" who established *Goals for Americans*.[37] Above all, the panic shunted aside the fiscal restraints imposed by the administration. Consistent with past efforts, it had begun an "economy drive" in 1957 to cut defense expenditures. Arms programs were canceled, reduced, or stretched out.[38] Before the panic, government intervention in the economy and in other aspects of life was kept to a minimum. After the start of the panic, Eisenhower was unsuccessful in his efforts to retain traditional limits on state action. *Sputnik* gave rise to four major initiatives: the doubling of research-and-development funds by 1961, massive federal aid to education, defense reorganization, and the space program.

While the president steadfastly denied the existence of a missile gap, he publicly worried that the Soviet Union had more scientists and engineers and was producing them at a faster rate. He appointed James Killian his special secretary for science and technology and elevated the Scientific Advisory Council to White House level. Edward Teller made a dire prediction: "Ten years ago there was no question where the best scientists in the world could be found—here in the U.S. . . . Ten years from now the best scientists in the world will be found in Russia."[39] Other scientists complained that the Soviets provided far more opportunity for basic research. The eduction system was found wanting: Russian children learned more science in a shorter time; science facilities in U.S. schools were inadequate, teachers poorly trained.

Financial restraint was now abandoned. Between 1957 and 1961, federal funds for research and development doubled, to $9 billion annually. Funding for basic research increased more than threefold, while the budget for the National Science Foundation more than doubled. More than $100 million was spent on a giant linear accelerator. Numerous high schools adopted the high-powered course in physics developed by the Physical Science Study Committee. However, the free flow of cash was not limited to physics and like fields. Americans set the goal of developing superior skills in every area of human endeavor. To that end, the National Defense Education Act of 1958 made available $250 million for public-school facilities.

The post-*Sputnik* atmosphere also encouraged fantastic fears. Ever since 1946, a pet project of the Joint Committee on Atomic Energy had been to develop an atomic airplane. Totally unsuccessful, the project was given a boost by the Soviet satellite. Fraudulent reports

37. See McDougall (1985, 217).
38. Huntington (1962, 95).
39. Cited in Kevles (1978, 284).

suggested that Russia was flying such a device. This ominous threat was perceived as yet another blow to the prestige and security of the nation and of the Free World.[40]

But the most enduring and consequential fear was that of a missile gap. In the immediate aftermath of *Sputnik*, Lyndon Johnson, the Senate majority leader, embarked on a major investigation of America's defense posture. A parade of witnesses from military, industrial, scientific, and educational institutions charged the administration with having ignored prior warnings; they made strident demands for new programs that were, in the words of General Curtis LeMay, "absolutely necessary."[41] Eisenhower's critics now found him vulnerable. Throughout the last years of his administration, the president was repeatedly labeled as complacent and constantly badgered for funds and programs to "catch up" with and move ahead of the Soviets. His efforts to assure the public that America's national security had not been compromised, that there was in effect no deterrent gap, fell on deaf ears. He was pestered with the missile gap even as he endlessly denied its existence. When he argued that augmented military spending would be inflationary and threaten national security by draining the economy, he was accused by the aspiring presidential candidate, John Kennedy, of seeking "economic security at the expense of military security."[42] Khrushchev did not help matters when he claimed that the Soviets were cranking out missiles "like sausages."[43]

But things grew worse still. In December, the *Washington Post* leaked the findings of the Presidential Commission on Civil Defense: "The still top-secret Gaither Report portrays a United States in the gravest danger in its history. . . . It shows an America exposed to an almost immediate threat from the missile-bristling Soviet Union."[44] The Gaither Report observed that although the Soviet gross national product was no more than one-third that of the United States, it was increasing at a much faster rate. Worse still, Soviet spending on defense and heavy industry was about equal to U.S. spending. The report warned that by late 1959, the Soviets could have 100 ICBMs with megaton nuclear warheads to attack the United States. Then came an even more startling revelation: America's Strategic Air Command bases were "soft" and vulnerable to a preemptive first strike.

40. Cited in York (1970, 70).
41. Cited in Clarfield and Wiecek (1984, 166).
42. Cited in Clarfield and Wiecek (1984, 165).
43. Cited in Beschloss (1986, 152).
44. Cited in Beschloss (1986, 149).

The focus on the capacity of the nuclear force to survive a first strike and then retaliate—effectively, a secure and survivable second-strike capacity—was to become a commanding issue over the next few years. The report recommended dispersing SAC bombers, constructing hard blast shelters to protect them, maintaining a significant number of bombers on "alert status," and constructing an early warning system to detect incoming missiles. It also recommended a $40 billion, five-year program of bomb-shelter construction and a dramatic increase in conventional forces. Finally, it proposed the deployment of 600 Titan and Atlas ICBMs, as well as 240 intermediate-range ballistic missiles at overseas bases.[45]

Eisenhower bristled at many of the recommendations. For all the brouhaha, the administration knew with a high degree of certainty that the Soviets had no operational ICBMs. This confidence was a result of photographs taken by U-2 spy planes, as well as electronic spying on Soviet test flights.[46] Since the Soviets had rejected an open-skies policy, the American leadership, seeing itself in a life-and-death struggle against an absolute enemy, made its own rules. The CIA developed and operated the U-2 spy plane, which flew at altitudes that Soviet interceptors could not reach. Aside from a few diplomatic protests, the Soviets never said anything about these incursions. They were not about to admit that the United States could overfly their country with impunity. Nor was Eisenhower ready to publicly disclose that the United States was secretly spying on the Soviets. The two sides were tacitly colluding to preserve the other's secrets. When the Soviets eventually downed a U-2 plane, publicly embarrassing the president, a State Department spokesperson read the following statement:

> Ever since Marshall Stalin shifted the policy of the Soviet Union from wartime cooperation to postwar conflict in 1946 . . . the world has lived in a state of apprehension about Soviet intentions. . . . I will say frankly that it is unacceptable that the Soviet political system should be given an opportunity to make secret preparations to face the Free World with the choice of abject surrender or nuclear destruction. The Government of the United States would be derelict . . . if it did not . . . take such measures as are possible unilaterally to lessen and to overcome

45. See Brown (1968); Clarfield and Wiecek (1984); Eisenhower (1965); and Kaplan (1983).
46. See Prados (1982).

this danger of surprise attack. In fact, the United States has not and does not shirk this responsibility.[47]

The U-2s found no evidence of a crash program to build and test ICBMs. By August 1958, the Soviets had launched only six ICBMs (four others failed). A year later, Allen Dulles, director of central intelligence, informed the president that the Soviets would probably have about ten missiles ready "either in 1959 or in 1960."[48] In September 1959, as the first American Atlas missiles were deployed, the U-2 photographs had not located a single operational Soviet ICBM. In January 1960, Dulles told Eisenhower that the Soviets still had no functioning ICBMs. Analysts estimated that they would have 35 by mid-1960 and between 140 and 200 the next year. By the end of the Eisenhower presidency in 1961, the only Soviet ICBMs were at two test sites.

Beneath the surface of these gaps lay a more interesting phenomenon. With its bombers in the 1950s and then its missiles in the early sixties, the Soviet Union did not embark on an all-out effort to develop the capacity to strike at the United States.[49] Rather, the Soviets concentrated initially on shorter-range missiles, for technical, military, and financial reasons. The first Soviet ICBMs were not only expensive but used unstable liquid fuel and still required ground stations for guidance. Concentration on shorter-range weapons made sense until these problems could be overcome, since medium-range systems fit the Soviet war-fighting strategy in the European theater. Effectively, the Soviets made the same decision as did Eisenhower: they deployed minimal numbers of first-generation liquid-fuel missiles to save funds for a second, solid-fuel generation. In effect, they delayed much of their ICBM program until 1964.

The Americans, by way of contrast, manufactured deficits from advantages and engaged in crash programs to gain and maintain a lead. Congress had panicked and, along with the public, created a stampede that Eisenhower termed "missile mania." Paranoia and mistrust, worst-case analysis, and faulty intelligence (one analyst complained that "to the American Air Force, every flyspeck on a film was a missile")[50] all helped fuel this sinister fear. The initial American ICBM plan had called for deploying 20 to 40 missiles. By 1956, 150 well-targeted missiles were thought to be enough; at no time before

47. Cited in Beschloss (1986, 257).
48. Cited in Beschloss (1986, 233).
49. See Holloway (1983, ch. 3) and Brune (1985, 21–25).
50. Cited in Prins (1982, 90).

1958 did the ballistic-missile group recommend more than 200 missiles.[51] But by the end of his term, Eisenhower approved a strategic-missile force that approached the 1,100 mark. According to Alain Enthoven and K. Wayne Smith:

> For a decade after Sputnik, however, the public mood in the United States was one of public support for almost anything proposed in the name of national security. During this period, the Secretary of Defense was under constant pressure to spend more money than he believed necessary. . . . The Armed Services Committee were rarely challenged by the rest of Congress.[52]

Calls for increased spending came from a wide range of interest groups, ready to capitalize on and direct the panic: the military, scientists, business associations, and organized labor.[53] According to Seymour Melman, a "state-management" decision system now supplanted the more privatized MIC of the mid-1950s.[54] Not surprisingly perhaps, in his farewell address Eisenhower took a parting shot at what he termed the "military-industrial complex."

But the missile gap would not go away. The Alsops wrote: "At the Pentagon, they shudder when they speak of the Gap, which means the years 1960, 1961, 1962 and 1963."[55] Indeed, it promoted the obsessive fear, especially among the civilian strategists, that the Soviets could launch a disarming preemptive strike against America's "soft" nuclear forces. The fear of a first strike, at a time when missiles were expensive prototypes, liquid fueled, highly unreliable, and extremely inaccurate, was simply unfounded.

As late as the 1960 presidential elections, however, the Republicans suffered at the polls for American somnambulism. John Kennedy made the missile and space gaps central concerns in the 1960 campaign. Despite Soviet overtures to control the arms race, the new president quickly launched an anticommunist campaign and presided over (what was then) the largest and most rapid military buildup in peacetime history.[56] Kennedy programmed the nuclear buildup *after* the missile gap had been officially debunked;[57] hence the decision

51. See Clarfield and Wiecek (1984, 172).
52. Enthoven and Smith (1971, 1).
53. Aliano (1975, 57).
54. Melman (1970).
55. Cited in Beschloss (1986, 150).
56. See Ball (1980); Kaufman (1964); Schlesinger (1965, 301); and Sorensen (1965).
57. See Ball (1980, 173).

cannot be attributed to faulty intelligence. Kennedy's electoral promises, which corresponded with a widespread desire for overwhelming nuclear supremacy, were certainly major factors in the buildup. The fear of a missile gap persisted, and speeches by Kennedy and Robert McNamara, secretary of defense, contain an inordinate number of comparisons and reassurances about American superiority. Pressure for supremacy came from the military, though its influence was probably not decisive.[58]

As noted previously, Eisenhower fought a rear guard action against increased government control and the technocratic threat to traditional American values. But the "emergency means" that he was compelled to put in place as a result of *Sputnik* became permanent. A U.S. technocracy was forged at the end of the Eisenhower years and then embraced and institutionalized by the new Kennedy administration.[59] Where state-funded and state-managed research and development seemed to contradict the tenets of a liberal economy, so great was the perceived threat to America's special destiny that the restoration of technological superiority was undertaken at the expense of these conventional values. In other words, some elements of the American way of life could be dispensed with in order to assure not only the survival of the United States but the continued belief in its historical and technological preeminence.

In 1958 the National Aeronautics and Space Administration (NASA) was established to oversee all nonmilitary activities in space. Congressman James Fulton made the agency's agenda clear:

> How much money would you need to make us even with Russia . . . and probably leap-frog them? I want to be the firstest with the mostest in space, and I just don't want to wait for years. How much money do we need to do it?[60]

The shock of *Sputnik* was compounded when the Russians launched an astronaut into earth orbit. Following the space flight of Yuri

58. Ball (1980) focuses on the effects of the MIC and downplays Kennedy's electoral commitments. Yet his evidence is often not consistent with his analysis. Besides observing that there was "little evidence of extensive industrial pressures" (254), he argues that McNamara originally planned for the procurement of 600 to 800 Minuteman missiles, while the military requested between 2,000 and 3,000 and, on occasion, 10,000 missiles. The final number of 1,000 is far closer to McNamara's figure. At the same time, the secretary of defense was able to cancel several major arms programs, including the Nike Zeus ABM and the B-70 bomber, and to impose systems analysis and other forms of civilian control on the military. See also Beard (1976).

59. McDougall (1985, 139–40).

60. Cited in Kevles (1978, 385).

Gagarin on 12 April 1961, which amounted to a second *Sputnik* shock, *International Affairs* (a Soviet journal published in English that now had a credulous audience) said the secret of Soviet spaceflight was

> rooted in the specific features of Socialist society, in its social structure, its planned economy, the abolition of exploitation of man by man, the absence of racial discrimination, in free labor and the released creative energies of peoples. Our achievements in the field of technology in general and in rocketry in particular are only a result of the Socialist nature of Soviet society.[61]

Not only America's special mission, but virtually every assumption of the civil religion is challenged here.

In his inaugural address, Kennedy had asserted that "we shall pay any price, bear any burden . . ."[62] Sacrifice now became a commanding theme of the administration. His immediate reply to the Gagarin flight was, "As I've said in the State of the Union Message, the news will get worse before it gets better, and it will be some time before we catch up."[63] On 25 May in a speech on urgent national needs billed as a second state-of-the-union address, he warned that "this nation is engaged in a long and exacting test of the future of freedom."[64] Kennedy's New Frontier was premised on perceived weakness, not strength. It was a call to action echoed by others. As Vice President Johnson, who was put in charge of the space program, observed, "In the eyes of the world, first in space means first, period; second in space is second in everything."[65]

President Kennedy quickly brought space into the superpower conflict when he called for the United States to land an astronaut on the moon before the end of the decade. The decision to go to the moon was the culmination of the new American technocracy. What had started as a military challenge was transformed into a civilian competition. In an ironic twist of Eisenhower's faith in the traditional

61. Cited in McDougall (1985, 245).
62. Cited in McDougall (1985, 308).
63. See Logsdam (1970, 102–4).
64. Cited in Logsdam (1970, 127).
65. Cited in McDougall (1985, 320). Secretary of State Dean Rusk wrote to the Senate Space Committee that "under the conditions existing in the world today, achievements in outer space have to assume a special significance in the assessments made by other countries of the strength, vitality and effectiveness of the United States and the free way of life we represent" (Logsdam 1970, 118).

American values of free enterprise and government nonintervention, Project Apollo was dubbed a peculiarly "American" enterprise.[66] In the new America of command technology and command innovation, the Apollo project, like the earlier Manhattan Project, became the model for further endeavors, particularly President Johnson's Great Society. As a result of Soviet challenges to sacred beliefs about the future of the nation, the United States sought to develop superior skills in every field of human activity, and technocracy became a primary means of doing so.[67]

66. See Mailer (1969). According to Logsdam (1970, 162), "The lunar landing decision followed from a definition of national interest that was in many ways peculiarly and transcendentally American. The attitudes that were at the root of the need to be first in space were more basic than just the necessity of staying ahead in Cold War politics. The desire to be successful, to make progress, to be preeminent is a powerful factor in American life. There is something close to the concept of 'national mission' in Kennedy's belief that 'whatever mankind must undertake, free men must fully share.' " True to tradition, Eisenhower lambasted the New Frontier and the Great Society in the *Saturday Evening Post*.

67. See McNeill (1982, 368).

7

The Cuban Missile Crisis

Sputnik notwithstanding, the late 1950s were characterized by reduced tensions between the superpowers. Khrushchev promoted the idea of peaceful coexistence, even as he declared that the USSR would bury the United States. Following a binge of atmospheric nuclear tests in 1958, November of that year saw the start of a test moratorium that lasted three years. Agreement on a comprehensive test-ban treaty appeared to be possible in early 1960, but was undermined when the Soviets shot down a U-2 spy plane. The most prominent issue still to be settled was Berlin. The Soviets used a combination of threat and cajolery to try to force an agreement. American concessions in this instance, the Soviets made it known, would produce a "thaw" in the Cold War. In January 1960, Khrushchev proposed an extensive reduction in Soviet forces, presumably in anticipation of a further relaxation of tensions. The Soviets let it be known that they might curtail their arms programs if the United States did not embark on further buildups.[1]

But superpower relationships deteriorated under the new Kennedy administration. Cold War rhetoric and confrontation replaced the faltering attempts at peaceful coexistence. Berlin, the Congo, and Cuba all became hot spots. Kennedy was determined to stop what he viewed as the insatiable imperialism of the Soviet state. Strong actions backed by superior military force became the call of the day. Kennedy inaugurated a massive military buildup, one that can be considered the culmination of the *Sputnik* panic and the associated missile gap. Part of the nuclear buildup involved a shift in strategy. In the place of massive retaliation, which seemed to be creating a stark choice between humiliation or an all-out war that verged on mutual suicide, the Kennedy policy was based on the doctrines of "flexible response" and "damage limitation." This meant avoiding cities and aiming instead at Soviet military installations (a counterforce strike), with the hope that this would induce the Soviets to spare American cities. The president, in effect, was trying a new strategy to escape paralysis and thereby render atomic weapons usable. Added to the nuclear strategy was a conventional forces buildup that reversed the steady reductions that had taken place under the Eisenhower administration. Kennedy feared that the Soviets would hide behind their missile power and use internal subversion and brushfire wars to nibble away at the periphery of the Free World.

The event that commands our attention is the Cuban missile crisis— the first, and thus far only, nuclear crisis. Although short-lived, it changed practically everything in superpower relations. More than any preceding event, it gave rise to a number of radical shifts characteristic of the nuclear world. Like *Sputnik,* the missile crisis challenged basic beliefs about America's special destiny. Like *Sputnik,* this third moral panic was a pure panic of omnipotence, portending the vaporization of American cities. In fact, the Cuban missile crisis can be considered the culmination of the "missile fear" unleashed by *Sputnik.* Where the launch of the satellite unbottled the demon of missile fear, five years later the Cuban missile crisis imploded that fear into a fortnight of unrelenting terror.

But having stared at the terror for two weeks, both sides recoiled. Thus the panic illustrates, better than anything else, the "paralysis" pole of the paradox of indispensability. In its aftermath, both sides seek new agreements, end nuclear brinkmanship, and scrupulously avoid superpower confrontations. But while Cold War tensions cool off, the arms race continues on its convulsive path, becoming un-

1. See Powaski (1987, 93).

hinged from diplomatic, political, and military needs. This disjunction is not surprising, since the missile crisis sanctified the belief that the Soviets could not be trusted. The surreptitious attempt to place nuclear missiles in Cuba solidified the view that the Soviets, in direct analogy to Satan, were purveyors of "totalitarian omnipotence." Corresponding with this position, there is an extension of the arms race from American superiority to the idea of exterminism, the perverse notion that only assured destruction or worse could deter the absolute enemy.

Finally, in the public culture, the missile crisis gives rise to the "great forgetting." The "profound nuclear fears"[2] that had convulsed the nation since 1949, and particularly in the late 1950s, were too intense to sustain. Now nuclear issues were submerged, and the public wanted only vague assurances (without the corresponding threats) that the American arsenal was sufficient to absolutely deter the Soviets. This chapter focuses on Cuba and civil religion, and Cuba and the paradox of indispensability; I will analyze nuclear forgetting the totalitarian omnipotence in Chapter 8.

Cuba and the Threat to America's Special Destiny

Analysis of photographs taken by a U-2 spy plane over Cuba on 14 October 1962 revealed installations under construction for medium- and intermediate-range ballistic missiles. Also being constructed were surface-to-air (SAM) installations. Whereas the latter weapons were inherently defensive, the former missiles could be armed with nuclear warheads and used offensively. When operational, they would leave a good part of the United States open to nuclear attack from Cuba.

The confidence of Americans in their own special destiny had taken a battering with *Sputnik*. The missile gap, Yuri Gagarin's space flight, and the abortive Bay of Pigs invasion all served to further attenuate American confidence. But the idea of enemy missiles ninety miles off Florida rendered the fear uncorked by *Sputnik* too near and too palpable. Ballistic missiles were still a relatively new and terrifying phenomenon, and *any* hope that Americans would prevail in the Cold War against their satanic foe required that they be capable of stopping

2. The phrase is used by Boyer (1980, 822).

them from introducing, surreptitiously, absolute weapons in America's backyard. That the missiles provided a degree of nuclear symmetry—the Soviets would now have a foreign base to match U.S. missiles installed in Turkey in the late 1950s—did not impress Americans.

The fear unleashed by the Cuban missiles was largely independent of their strategic value. As Kennedy asserted, the Cuban bases did not fundamentally alter the nuclear balance. They did not afford the Soviets any possibility of a first-strike capability, and the need to fire from the Soviet Union and Cuba simultaneously resulted in a "ragged volley." The Cuban installations were "soft," vulnerable to a counterforce strike. Since the missiles were liquid-fueled and took several hours to fire, they made poor second-strike weapons. The Cuban force did, however, complicate a possible American first strike, which the air force now thought feasible.[3]

From the Americans' point of view, they had been confronted with Soviet duplicity. Anatoly Dobrynin, Soviet ambassador to the United States, had given a special message to presidential aid Theodore Sorensen. He was to tell the president that "nothing will be undertaken before the American Congressional national elections that would complicate the international situation."[4] There had also been assurances that no offensive weapons were being installed in Cuba. The U-2 photographs revealed that both statements were patently untrue. The missiles, while not altering the strategic balance, upset the status quo too much. Any act that significantly increased the chance of using such weapons or suddenly changed their pattern of deployment would be regarded as extremely menacing in an already sensitive nuclear world. The absolute weapon fostered a sense of omnipotence but also put it at perpetual peril.

The existence of a communist regime, armed by the USSR, in the Western Hemisphere ran headlong into the Monroe Doctrine. The United States, like the Soviet Union, was unwilling to tolerate a hostile or ideologically foreign regime in its own backyard. To add to the insult, Cuba was now openly embracing the Soviets. The addition of the missiles made it intolerable. As Robert Kennedy was to say to the president, if he did nothing about the missiles, he would be impeached.[5] To the president, the missiles were less important strategically than politically; in his words, the missile deployment "politically changed the balance of power. It . . . appeared to, and appearances

3. See Prins (1982, 92–93).
4. Cited in Detzer (1979, 52).
5. See Kennedy (1969).

contribute to reality."[6] The pride of the administration *and* the pride of the country were at stake. The president had seemed uncertain in foreign affairs, especially after the Bay of Pigs fiasco. He needed a victory to salvage his faltering image. There was a need to take a firm stand, to reassert American (and presidential) strength in the face of Soviet challenges.

To people all over the globe, Kennedy's decision to impose a limited quarantine on Cuba brought home the reality of thermonuclear war. Many believed that the end was at hand; that the unthinkable was about to happen. Peoples and nation-states had become hostages to the arsenals of the superpowers. At the United Nations, African and Asian politicians expressed the realistic fear that the political objectives and pride of the United States were bringing the world to the brink. These leaders objected to their status as hostages and, quite possibly, victims. They were angry with Kennedy and pressed U Thant, acting secretary general, to intervene.

Reaction in Britain was far from positive as well. The *Daily Telegraph* called the blockade "greatly mistimed."[7] The *Guardian* observed that "what is sauce for Cuba, is also sauce for Turkey." The *Daily Mail* held that "President Kennedy may have been led more by popular emotion than by calm statesmanship."

The latter point bears closer examination. The crisis engendered tremendous fear in the United States. People snapped up civil-defense leaflets, bought sandbags and other blast-resistant materials, and simply prepared for the worst. Yet for all the alarm, there was remarkably little dissent and tremendous support for Kennedy's actions. Opposition came from some college campuses—the *Harvard Crimson* accused Kennedy of a "frenzied rejection" of diplomacy.[8] But at Stanford the student newspaper wanted to "Invade Cuba." Anti-quarantine demonstrations were held, but the demonstrators were typically outnumbered by those favoring action.

A Gallup poll taken 23 October revealed that 84 percent (of those who knew about the situation) favored the blockade; only 4 percent opposed it.[9] This in spite of the fact that the same poll revealed that one out of five Americans thought that it would lead to World War III. Public support may have been stronger than the polls revealed! Gallup's respondents frequently said things like, "It should have been

6. Cited in Brown (1968, 259).
7. Citations in Detzer (1979, 203).
8. Citations in Detzer (1979, 191).
9. See Detzer (1979, 192).

done a lot sooner" or "It was long overdue." Other common comments were "We've been pushed around long enough" and "We had to draw the line somewhere."

A crowd of more than eight thousand Conservative party members at Madison Square Garden booed the Kennedy announcement. The quarantine was "too little, too late." Speakers demanded an immediate invasion of Cuba, and delegates shouted "Fight! Fight!"[10] Republican congressmen also favored a more vigorous response. Critics thought that Kennedy had missed a golden opportunity, achieving only the "bare minimum"[11] by getting the missiles removed. There was still a communist regime in the Western Hemisphere that threatened the American way of life. For many, the noninvasion pledge amounted to the surrender of U.S. interests and power. Daniel Moynihan termed the agreement a "defeat."[12] Kennedy should have forced the Russians out altogether. He might very well have overthrown the Castro regime. Many felt betrayed.

One thing above all seems clear. The placing of offensive missiles in Cuba was intolerable to both the American political leadership and its public. The weapons had to be eliminated or removed. On this, Kennedy had near-unanimous support, which would have backed virtually any military action that he might have undertaken. Robert Kennedy's remark that his brother would have been impeached if he had done nothing was correct not in fact but in manner of speaking.

United States prestige (Would the Americans allow the Russians to get away with this in front of the whole world?); the electoral fortunes of the Kennedy administration; and a strategic threat with more political than military ramifications: these were all implicated in the crisis. But underlying them all was a moral panic of omnipotence with a heritage that extended back to the explosion of the Soviet atomic bomb and forward to *Sputnik*. As in the latter two events, the greatest panic and the most concerted public and political responses devolved around sudden jumps in Soviet nuclear capabilities. The proximity of the Soviets in Cuba had already sparked pressure to do something. An incipient panic was under way even before Kennedy made his announcement. But the missiles were the terrifying and unacceptable culmination of the extension of Satan's duplicitous power into America's backyard. They challenged U.S. superiority, practically and symbolically. If American vigilance was not sufficient to deter Soviet trickery, if the nation's power was not great enough to stop the Soviets

10. Citations in Detzer (1979, 187).
11. David Lowenthal, cited in Divine (1971, 92).
12. Cited in Detzer (1979, 257).

from introducing absolute weapons into its backyard, then what remained of American security, of American moral certitude, and of America's special destiny? The threat that the absolute enemy might prevail in this particular instance was so insufferable that Americans could even countenance going to the brink.

Even before Kennedy was assassinated, the Cuban missile crisis was seen as his finest moment. Fury had taken precedence over reasoned calculations in the popular mind. The Cuban parallel with Turkey, so clear to much of the rest of the world, had little impact on Americans. A swap of the two sets of missiles was unacceptable for many reasons; certainly among these was the sense that a trade, detestable in itself, would serve to equate Soviet and American actions and intentions.

The fact that the crisis itself was only an episode, and that in the next few years the Soviets would deploy large numbers of more reliable ICBMs that would have rendered the Cuban missiles obsolete and superfluous, was also unimpressive. I. F. Stone, an independent journalist, had earlier warned of such overreactions to transient crises:

> There is no presumption more terrifying than that of those who would blow up the world on the basis of their personal judgment of a transient situation. I do not propose to let the future of the world be settled, or ended, by a group of men operating on the basis of limited perspectives and short run calculations.[13]

But the broader moral panic that gripped Americans belied attempts to cast the events into a personal or even partisan political mold. The Kennedy administration played out the game with great skill. Its members did so in a political context which demanded that they engage in it. Andrei Gromyko, the Soviet foreign minister, was correct when he warned Khrushchev that deployment of missiles in Cuba would trigger a "political explosion" in the United States.[14]

Cuba and the Paradox of Indispensability

Since it remains the only nuclear confrontation between the super-powers, the Cuban missile crisis has been analyzed ad nauseam. Here

13. Cited in Divine (1971, 161).
14. Allyn et al. (1989–90, 148).

the concern is not with whether Kennedy made the best moves in 1962 or the extent to which the affair is relevant to the management of a current crisis, given subsequent advances in nuclear capabilities. Rather, I argue that it was the definitive lesson in nuclear paralysis, especially from the political point of view. A twenty-fifth-anniversary meeting of former policy makers from the Kennedy administration and scholars of the missile crisis in Hawk's Cay, Florida, in March 1987 presented evidence supporting this conclusion. Specifically, the evidence is consistent with the claim that "President Kennedy had decided he was not going to war over the missiles in Cuba, but that he would do his utmost to get them removed with the least political cost."[15]

Here I show the extraordinary lengths to which Kennedy and Khrushchev went in order to avoid an escalation of the crisis. Implicit in their behavior is a model of deterrence that does *not* rely on the credibility of a threat but on the fear of inadvertent escalation or eruption. The combination of panic and paralysis is critical for understanding the effects of the crisis on the nuclear world.

Altogether, the development of the missile crisis can best be regarded as a study in escalation avoidance. American proposals were tested against a three-runged escalation ladder: What would the Soviets do in response? What do we do then? What do they do then?[16] At virtually every step, Kennedy took the decision that would minimize the confrontation, lessen the chance of escalation, and afford the Soviets the most room in which to maneuver.

Since the idea of tolerating the missiles was unacceptable, while the risk of allowing them to become operational was too dangerous, the Americans opted for the minimum military option: a quarantine of Cuba limited to offensive weapons. Whereas Under Secretary of State George Ball observed that "as far as the American people are concerned, action means military action, period,"[17] the Executive Committee (ExCom) of the National Security Council, set up to advise the president, realized that it could not force the Soviet hand. While several leading congressmen who were forewarned about the blockade thought it a half-measure and advocated instead a full-scale invasion of Cuba, those involved in the decision making opted for low-keyed actions.

15. Blight, Nye, and Welch (1987, 180) base their conclusion on evidence given at Hawk's Cay.

16. Chess grand masters think at most three moves ahead. Escalation ladders with up to twenty or more steps are nice academic exercises with no correspondence in reality.

17. Cited in Brune (1985, 47).

One major option was an air strike, but this "fast track" proposal contained the real possibility of escalation. In particular, there was the fear that the Soviets would retaliate somewhere else, where the correlation of forces was more in their favor. During a discussion of the possibility of a Soviet attack on U.S. missile bases in Turkey, the president asked an unidentified "hawk" how the United States would respond. The reply was: "Under our NATO treaty, we'd be obligated to knock out a base in the Soviet Union." Kennedy asked again, "What will they do then?" The response was: "Why then we hope everyone will cool down and want to talk."[18]

The blockade (termed the "slow track"), in contrast, was a prudent military option. It avoided a direct attack on Cuban soil and resultant Soviet casualties. And unlike the air strike, it was not irreversible. Of course, it might not work. Indeed, it allowed the Soviets time to render the missiles operational.

In his nervous announcement of the blockade, Kennedy warned that a nuclear attack on any part of the Western Hemisphere would be taken as a direct attack by the Soviet Union on the United States and would lead to a full retaliatory response. In effect, the president was denying that there was an escalation ladder with more than a few rungs and that any possibility existed for limiting nuclear war. Perhaps Kennedy threatened full retaliation in order to put any nuclear exchange beyond the pale of the possible. Nuclear bargaining and the testing of resolve can be analyzed theoretically, but they have no place in the epicenter of a crisis.

The blockade was accompanied by a series of diplomatic moves. Support was sought and gained from the Organization of American States and from the United Nations. Emissaries were sent to U.S. allies before the blockade announcement. Contact was also established with the Soviets. Robert Kennedy spoke to Ambassador Dobrynin, while the president sent a rather conciliatory note to Khrushchev. He asked the Soviet leader to act prudently and not to "allow events to make the situation more difficult to control than it is."[19]

The blockade was managed so as to minimize the chances of eruption. The quarantine itself was set up five hundred miles from Cuba, far enough so that Russian MIGs based there could not reach it. As Soviet ships approached, Kennedy ordered the blockade to contract in order to allow them more time. Strict orders were given stating that authorization to fire had to come from the White House.

18. Citation in Brune (1985, 49).
19. Cited in Brune (1985, 61).

When it was confirmed that some Soviet ships had turned back, Kennedy acted to ease the pressure and to help Khrushchev save face. Thus he allowed the first Russian ship through the blockade. There had been some debate on whether or not to stop it. Those in favor thought halting it would display American determination. The opposition wanted to give Khrushchev more time. Kennedy, who tried to take his counterpart's point of view during the crisis, argued: "We don't want to push him to a precipitous action—give him time to consider. I don't want to put him in a corner from which he cannot escape."[20] Significantly, when the tanker *Bucharest* was contacted by the American fleet, it acknowledged the call and declared that its cargo was petroleum. Despite the rhetoric—Khrushchev initially stated that the Russian ships would not respect the blockade—the Soviets were cooperating.

Since the Americans had to stop a ship to establish the credibility of the quarantine, they decided on the *Marcula*, an oil freighter built in the United States, chartered by a Russian organization, and flying a neutral flag. This action would be less irritating than stopping a Soviet or East European vessel. The boarding party wore whites and carried no weapons. Both sides had worked to avoid a confrontation.

A nuclear crisis was unprecedented, and its resolution also came from unprecedented means: low-level contacts outside normal diplomatic channels and personal letters between the two leaders. Proposed solutions were passed between John Scali, diplomatic correspondent of the American Broadcasting Company, and Alexander Fomin, officially listed as a counselor at the Russian embassy, who had approached Scali in the first place. Presumably, these contacts were used to float trial balloons that could be officially denied. More significant, perhaps, were the personal letters sent between the two leaders. Indeed, nothing better expresses the paralytic fear they faced than Khrushchev's rambling letter of 26 October, which included his "knot" metaphor:

> If you have not lost your self-control and sensibly conceive what this might lead to, then, Mr. President, we and you ought not now to pull on the ends of the rope in which you have tied the knot of war, because the more we pull, the tighter the knot will be tied. And a moment may come when the knot will be tied so tight that even he who tied it will not have the strength to untie it.[21]

20. Cited in Detzer (1979, 228).
21. Cited in Abel (1966, 183).

Instructive, too, was the American response to a U-2 plane being shot down by a SAM missile over Cuba on 27 October. Several days earlier, the ExCom agreed that if a U-2 were fired on, American planes would retaliate by destroying a SAM site. According to Robert Kennedy, the ExCom "almost unanimously" agreed upon a strike on Cuba when the American plane was shot down. The Joint Chiefs of Staff also called for the destruction of the offensive-missile installations. Kennedy, however, exercised restraint.[22] He decided to send another reconnaissance flight the next day to test Russian intentions. Only if it were fired on would the Americans counterattack. On the same day the U-2 was shot down, another U-2 on a weather flight strayed over Soviet air space, giving credence to fear of inadvertent escalation from below. According to Kennedy, who had previously ordered all such flights canceled, "There is always some son-of-a-bitch who doesn't get the word."

The president, of course, prepared for war. Robert Kennedy sent signals that could be interpreted as an ultimatum.[23] However, recent evidence sent to the Hawk's Cay conference by Dean Rusk indicates that in the final analysis John Kennedy would have traded the Turkish missiles rather than have risked war. According to the former secretary of state:

> There is a postscript which only I can furnish. It was clear to me that President Kennedy would not have let the Jupiters in Turkey become an obstacle to the removal of the missile sites in Cuba because the Jupiters were coming out in any event. He instructed me to telephone the late Andrew Cordier, then at Columbia University, and dictate to him a statement which would be made by U Thant, secretary general of the United Nations, proposing the removal of both the Jupiters and the missiles in Cuba. Mr. Cordier was to put that statement in the hands of U Thant only after a further signal from us. That step was never taken and the statement I furnished to Mr. Cordier has never seen the light of day. So far as I know President Kennedy, Andrew Cordier and I were the only ones who knew of this particular step.[24]

22. Allyn et al. (1989–90, 159) observe that at the height of the crisis the ExCom had become largely irrelevant to the president's decision making.

23. See Allyn et al. (1989–90, 163–65).

24. J. Anthony Lukacs, "JFK's Warriors Relive Cuban Missile Crisis," *Toronto Star*, 30 August 1987, H-5. See also Blight, Nye, and Welch (1987).

Revelation of the "Cordier maneuver" clearly indicates that the president was willing to suffer political rather than nuclear fallout.

As a final agreement was reached, Kennedy took care to avoid any suggestion of Russian "capitulation." Cooperation between the two sides expanded during the weeks required to resolve all the outstanding problems. Castro, no mere Russian puppet, did what he could to sabotage the deal. He never allowed on-site inspection of the dismantling process. In the end, the navy did a visual inspection of Soviet ships leaving Cuba. Castro claimed that the IL-28 bombers based on the island were a gift and not to be returned, while Khrushchev noted that they were not part of the original bargain. The United States kept up pressure, letting it be known that it might bomb the planes and that the pledge not to invade Cuba was valid only so long as all offensive weapons were permanently removed from the island. Castro relented. In the following year the United States unceremoniously removed its Jupiter missiles from Turkey.

8

Nuclear Forgetting versus Totalitarian Omnipotence

A sign on the door of a conference room in the State Department read: "In a Nuclear Age, Nations Must Make War Like Porcupines Make Love—Carefully."[1] The sign was not sufficiently radical. In the course of the missile crisis, the leaders of both sides discovered that they could not make war at all. The crisis had revealed that the superpowers were, in a metaphor suggested by Oppenheimer, like two scorpions in a bottle, each capable of killing the other but only at the risk of its own life. Atomic weapons had "spoiled" war, a conclusion recognized in the first book on nuclear strategy, *The Absolute Weapon,* published in 1946 when only the A-bomb, in very limited numbers, existed. According to Frederick Dunn: "In fact to speak of it as just another weapon was highly misleading. It was a revolutionary development which altered the basic character of war itself."[2]

1. Cited in Detzer (1979, 131).
2. Dunn (1946, 4).

The scorpion dilemma was certainly the most obvious and enduring lesson of the missile crisis. For all their differences, the superpowers were bound together by one profound interest: avoiding nuclear war. The weight of the missile crisis made its presence felt in the Cold War. The terror drove them to détente—to a reduction of tensions and to a degree of peaceful coexistence that allowed them to "settle" the points of contention in Europe. Russia had consolidated its hold on Eastern Europe during the period of the American atomic monopoly. In the aftermath of the missile crisis, it was clear that, beyond diplomacy, nothing could be done to change the European configuration. Khrushchev had once observed that Berlin was the testicles of the West. Every time he squeezed, they hollered. Now the period of squeezing was over.[3]

To use a term not in currency at the time, the missile crisis opened an extraordinary "window of opportunity" between the two superpowers. Beyond détente, the two powers now scrupulously avoided both confrontations and nuclear brinkmanship. The crisis also led to some arms agreements. For years, peace groups had been protesting the poisoning of the atmosphere by nuclear tests. In 1963 the two nations signed a partial test-ban treaty stopping atmospheric nuclear testing (the agreement did not ban underground testing). The irony is that the two sides came within three inspections of a total test-ban treaty. Where the Soviets agreed to four annual on-site inspections, the Joint Chiefs of Staff argued that a minimum of seven such inspections was required to verify Soviet compliance. The joint chiefs prevailed and thereby blocked what may have been the most important "missed opportunity" since Trinity.[4] A total test ban would have effectively ended the arms race, since neither side would rest its fate on untested warheads. The two countries also established a "hot line" in order to facilitate communications in a future crisis. Subsequent agreements included banning weapons of mass destruction in outer space and signing a nuclear nonproliferation treaty.

In the public culture the window of opportunity comes as "nuclear forgetting," a concerted effort to forget, ignore, or suppress, by whatever means necessary, nuclear fear. In defining nuclearism, Robert Lifton suggests that

3. Détente would be derailed in the mid-1970s, but the superpower confrontations never again reached the pre-détente level. Reagan's encounters with the Soviets were mainly in the realm of rhetoric, as well as in the ongoing arms race.

4. The Joint Chiefs of Staff opposed a total test ban since they wanted to continue to build and test new warheads. York (1970) argues that the joint chiefs, with their allies in the Senate, would have vetoed anything more than a partial test ban. A tacit understanding on these matters may well have been reached with Kennedy.

we seek in the dazzling performances of technology and science a replacement for something missing in our lives. We invoke the illusions of the "bomber gap" or "missile gap" in order to fill the meaning gap. In what may be the ultimate human irony, we seek in a technology of annihilation a source of vitality, of sustained human connectedness or symbolic immortality. At the heart of this illusory structure is our struggle for a sense of power, which in the end turns out to be power over that ultimate adversary. The conquest of death—the restoring of youth to aged—is after all a central goal of premodern alchemy.[5]

But in the aftermath of the nuclear crises, there is an important shift in the culture surrounding nuclearism. The "nuclear obsession" that prevailed before 1963 is now drained of much of its intellectual and emotional force. The bomb, as Paul Boyer observes, did not totally vanish from the American consciousness.[6] But the country turned its attention elsewhere; instead of the bomb occupying center stage, Americans wanted vague reassurances about their nuclear superiority (a critical issue I discuss below) and otherwise to see and hear virtually nothing about nuclear realities. Indeed, the orthodox faith in nuclearism was beginning to give way to nonconformity.

This decline of nuclear awareness and activism can be explained in a number of ways. For Lifton, the key factor is psychic numbing, a relatively straightforward psychological process. Boyer, in contrast, looks to more specific historical changes, especially those that create a cultural climate conducive to apathy and complacency.[7] Thus he focuses on the following factors: the perception of diminished risk, a result of the thaw in the Cold War; the loss of immediacy, a result of familiarity, which can take the sting out of anything, Armageddon included; the shift to underground testing, which removed the visible and pressing danger of radiation; the neutralizing effect of the peaceful atom; the complexity and reassurance of nuclear strategy, a result of the shift to deterrence and the end of brinkmanship; and, finally, the Vietnam War and the rise of the New Left.

While Boyer recognizes the profound nuclear fears that had

5. Lifton and Falk (1982, 95).
6. Boyer (1980); this article is central to the present discussion. Evidence from public-opinion polls is also consistent with the Great Forgetting. Thus McClelland (1977) finds that the "Cold War pattern" ends by 1964, and other issues emerge as primary concerns. See also Abolfathi (1980).
7. See Boyer (1980).

gripped Americans, I think his analysis underestimates the role of panics in creating cultural conditions conducive to specific historical changes. In other words, there was a powerful interaction between historical changes and the psychological and cultural processes embodied in moral panic. Where the "panic and reaction" cycle previously led to nuclear buildups to restore the threatened sense of American omnipotence, the two decades of nuclear fear that were brought to a head by the missile crisis culminated in a cultural shift in which nuclear concerns were forsaken.

Consider the partial test-ban treaty of 1963. Boyer suggests that the test ban was a critical factor in the loss of immediacy, and I agree. However, he ignores the moral panics that rendered the test ban almost inevitable. In his view, the test ban "won enthusiastic public and journalistic support."[8] It gave rise to a sense of euphoria, to the belief—this sentiment reoccurs near the end of 1988!—that "Peace has broken out, and hope leaps up again." A Harris public-opinion poll revealed that unqualified approval of the treaty reached 81 percent by 1 September 1963.

But such enthusiastic support for a test ban did not exist between 1954 and 1958. A detailed content analysis of fifty periodicals and the *New York Times* during those years reveals a contested debate.[9] The magazines split almost evenly in supporting or opposing the test ban. Most clustered in the middle, with the *New York Times* taking a conservative anti-ban position until 1958. Indeed, support for the ban peaked in 1957; opposition was strongest in 1958—an immediate result of *Sputnik,* perhaps? On the whole, mass-circulation magazines did not favor the ban, while support for it was strongest in the limited-circulation "quality" journals. The amount of interest shown in the issue and the quality of the intellectual debate were greater in the latter periodicals.

That the test-ban treaty won such enthusiastic support in 1963 cannot be attributed simply to the debate itself. Rather, I am suggesting that the cumulative effect of the moral panics, and especially the missile crisis, was to create a psychological and cultural climate that demanded a diminishing of the (visible) nuclear menace. Fear, like any emotion, is self-limiting. Americans had been strained by such profound fears that they were willing to forge an agreement with the Soviets despite what Glenn Seaborg, chairman of the AEC, termed the main obstacle: the fear of Soviet cheating. This concern was

8. Boyer (1980, 821).
9. See Rosi (1964).

probably the main reason that an agreement on a total test ban could not be reached. A partial test ban removed the immediate fear of radiation poisoning without rendering Americans susceptible to Soviet cheating.

The argument that moral panics created psychological and cultural conditions conducive to specific historical changes can also be applied to the détente that followed the missile crisis. Unlike the thaw after the death of Stalin, détente was not a result of significant changes in Soviet behavior. Rather, as in World War II, when the countries were allies, Americans were simply willing to disregard, because of other interests, oppressive aspects of Soviet behavior. The new guiding interest was the avoidance of superpower conflicts, of any confrontation that could possibly escalate to a nuclear level. The moral panics that forged nuclear forgetting created a public reality in which such a conflict was impermissible. The Cold War and every other issue were subordinated to this primary concern. Where Boyer suggests that the Great Forgetting gave rise to apathy and complacency, I maintain that the fear was still there, always hovering just below the surface. Americans were less apathetic than willing to do whatever seemed necessary to keep the fear at bay, to avoid situations that could stoke it.

Here we come to a critical and vexing disparity: the unfolding of the arms race in the post–missile crisis world. Despite the hope and the seeming window of opportunity, the partial test-ban treaty did not turn out to be the first step in a process of reducing the nuclear menace. The panics did not undermine the indispensability of these *now* clearly unusable weapons. Their profound mutuality of interests did not drive the two sides to significant arms-limitation agreements. Tacit understanding, common enough in the Cold War context, did not materialize in the nuclear realm. The balm of political thaws did not extend to the arms race; the two were uncoupled and would remain so even as political and diplomatic relations improved.

After the assassination of President Kennedy on 22 November 1963, Khrushchev quickly impressed on the new president the Soviet desire to sustain détente. On 13 December he announced that the USSR intended to reduce its military budget and was considering the resumption of troop reductions that had been stalled by the Berlin crisis. A 5 percent reduction in Soviet military expenditures was paralleled by a $500 million cut that President Johnson later announced in the defense budget. Khrushchev hinted that progress in arms reduction could be achieved by a process of "mutual example."[10]

10. See Horelick and Rush (1965, 155).

In the absence of formal agreements and open inspection procedures, each side might reciprocate the arms-reduction measures undertaken by the other. This did not occur. Nor did President Johnson's freeze proposal in early 1964 have any effect. The United States and the Soviet Union continued with the deployment plans that had been set out before the crisis, and soon augmented even these.

In the five years following the test ban, both sides conducted more nuclear tests than in the previous five years. At the same time, nuclear weapons and delivery systems were expanded at an unprecedented rate. Whereas the United States possessed about 300 ICBMs (compared to perhaps as many as 75 for the Soviet Union in 1962), by 1969 each side had about 1,050 long-range missiles.[11] Where submarine-launched ballistic missiles (SLBMs) were just being deployed in 1962, by 1969 the American submarine force consisted of 656 SLBMs, while the Soviets had close to 200 and were rapidly deploying more. The United States still had more than 500 strategic bombers and the Soviets about 150. The Soviets also had a large number of medium- and intermediate-range ballistic missiles in the European theater. Both sides also deployed a wide range of smaller, tactical nuclear weapons meant to be used on the battlefield. Most significantly, in 1964 Johnson equipped Polaris SLBMs with multiple warheads and programmed the Minuteman III and Poseidon missiles that would be equipped with multiple, independently targeted reentry vehicles (MIRVs). MIRVs were initially developed to counter a feared Soviet antiballistic missile (ABM) system, and would prove to be extremely destabilizing.[12]

In 1967, Dean Rusk, the secretary of state, observed: "Our military power . . . is so vast that the effects of it are beyond the comprehension of man. It is so vast that we dare not allow ourselves to become infuriated."[13] At about the same time, McNamara declared that the U.S. buildup was "both greater than we had originally planned and in fact more than we require."[14] Indeed, since "flexible response" failed its first test in the missile crisis, McNamara subsequently opted for a policy of "assured destruction." But since the Soviets undertook a massive nuclear buildup of their own, assured destruction became mutual assured destruction or, fittingly, MAD. By 1969, exterminism was fully in place.

11. The Soviets may have had as few as twenty missiles deployed in 1962. See Allyn et al. (1989–90, 141–43).
12. This is discussed in Chapter 9.
13. Cited in Moss (1968, 330).
14. Cited in Ball (1980, xxiv).

The Bomb and the Creation of Totalitarian Omnipotence

The obvious question in this context is why the Cuban missile crisis contributed to the easing of political tensions but not to the easing of the arms race. What was driving the arms race forward? Or, what was rendering it so difficult to control? Here I seek to show that the bomb, as a confounding power, was itself a critical factor in the arms race. The key to its role is not to be found in deterrence as such. Rather, to repeat what P.M.S. Blackett observed, "When a nation pledges its safety to an absolute weapon, it becomes emotionally essential to believe in an absolute enemy." When the bomb's confounding power and Soviet totalitarianism were fused, they gave rise to the terror of totalitarian omnipotence. The attribution of such perversity to the Soviets is so consequential that it merits systematic analysis of the general fear of communism.

Marxism is one variant of the utopian visions of progress so prevalent in the nineteenth century. The Bolshevik Revolution, coming near the end of World War I, fostered great hope among many liberal and leftist intellectuals in the West. But the revolution quickly turned into an iron cage, a repressive order shorn of humanitarian ideals. Progress, stripped bare, was again replaced by omnipotence. Faith was placed in human powers, as embodied in the Party and the state. In this case, the domination of nature was secondary to the domination of people. The requirements of the state, as formulated in the increasingly problematic "laws of history," took precedence over the individual. Totalitarian omnipotence—it was not *fully* realized until the Soviets had the bomb—removed all restraint in the submissive use and misuse of humans in the achievement of the inevitable ends of history. The present was mortgaged by the use of raw power for the ostensible achievement of some ideal condition in the indefinite future.

Totalitarian omnipotence is evil in a special way. An evil that is based on fury and anger is at least understandable. An outrage or atrocity that expresses an emotional upheaval is at bottom "motivated." It suggests that the passions have taken precedence over reason. Thus in Dante's *Inferno* the shades endure a hell commensurate with their own depraved passions. In totalitarian evil, however, fury gives way to simple cruelty, and anger is reduced to cold malice. Reason takes precedence over the passions. But it is a reason gone awry, a reason corrupted by pride and untutored by any limits. It is this godlike pride that characterizes the three-headed Lucifer in the bottom pit of the inferno. The pride that leads to the false ascension to the Godhead—the pride of Satan—is the definitive evil.

Totalitarian omnipotence brings Adolf Eichmann as its ideal type. (No Soviet example, including Stalin or Beria, appears to be as well known.)[15] Cool, neat, always polite, seldom given to emotional outbursts or displays, he was the systematic, efficient, and thorough assassin, smiling benignly as he sent death-pale children to the gas chambers. Accused by those who managed to evade his murdersome machine, he showed no remorse but defended and justified his actions. His one regret was that the Final Solution was not quite that. None of the empathy possible in the case of emotionally commanded evil is felt here. If only he were irrational with fury! It is difficult to express moral outrage in the face of such calculated malice. One is aghast and bewildered. It seems so incomprehensible. It is also terrifying. The fear of humanity has replaced the fear of God. It is the fear of a humanity swollen with the pride of its own godliness.

In totalitarian omnipotence the state is invested with the power to kill for the higher good. Genocide is used to shape the future; it becomes a tool to realize the New Order. Under the Nazi regime, doctors were recruited as "biological soldiers."[16] The death camps provided a great temptation, the power to cure the Nordic race, to prune its genetic stock according to the "laws of biology." In this applied biology, killing was healing. The Soviet state also accorded itself the right of logical murder. Here killing was based not on race but on class. Class origins were treated as akin to a hereditary affliction. Class enemies, as well as other enemies of socialism—and here Stalin's paranoia cast a wide net—were subject to physical and psychological brutalization, forced relocations that often separated parents and children, the Gulag, and, of course, extermination.

The exacting cruelty and disquieting domination of the Soviet regime, which was so bold as to stake a claim to the future of humankind, was central to its being cast as an absolute evil in Ameri-

15. There is a terrible irony here, as the Soviets detest, with good reason, the Nazi regime. Yet Western analyses often see a convergence between the totalitarianism of the Left and of the Right. Stalin's atrocities are certainly comparable with those of Hitler. The effort to control and engineer human behavior seems to have been carried further in communist regimes. The fear of a German atomic bomb was soon replaced by the fear of a Soviet one. Popular literature, in the form of spy and military stories, also relies on both of these systems. In each case, the villains seem to fit the ideal type depicted here: emotionless, almost robotic assassins with total faith and absolute cunning on their side.

16. The term is used by Robert Lifton (1986).

can eyes. One of the most enduring memories of the Korean War was the "brainwashing" of American prisoners of war by Chinese communists. Mind control meshed nicely with the fear of communist espionage and subversion. Communist spies and subversives were obviously brainwashed, which meant that they were infinitely dangerous. A cold cunning, unfettered by morality or emotion, as well as magical and seemingly omnipotent techniques for controlling the behavior of others, were at their disposal. Against such fiendishness, democratic societies could not help but be vulnerable. The Soviets were, in a recent phrase, "engineers of human souls." In the control of behavior and the suppression of dissent, compared to them even the Nazis were mere apprentices.

When the domination of humanity was coupled with the bomb, totalitarian omnipotence was fully realized and insecurity reached an unprecedented high. American omnipotence was vested in the country's awe-inspiring technology. But this technological hubris was vulnerable: there were atomic secrets, and Soviet agents were after them. If the possibility of the secrets being disclosed bred fear and insecurity, the explosion of Joe 1, the USSR's first atomic bomb, produced a quantum leap in these feelings. The Soviets had caught up in the realm of technological omnipotence. They now represented a portentous danger, for it was believed that the fear of this weapon, which (ostensibly) limited American behavior, would not constrain them. The calculating atrocities committed against the populations of their own and neighboring states were taken to indicate that they would use the bomb in a similar fashion. Soviet efforts to establish power over humanity admitted no exceptions; nuclear weapons would be employed, without scruples, to further their objective of domination. The fusion of the bomb and totalitarianism spawned, in a modern incarnation, Satan.

Metaphor is meant to be suggestive. However, the line separating it from "reality" is often fuzzy. Yet as Garrett Hardin, the noted ecologist, observes, we cannot think without the use of metaphor. The Soviet-Satan connection is metaphor—and yet more. That is, in the course of the moral panics, the capacity to draw distinctions between metaphor and reality was powerfully attenuated. I will flesh out the Soviet-Satan connections, and then show how closely American statements about Soviets and the bomb fit them.

Ubique daemon—the devil is everywhere. Perhaps no other statement so succinctly encapsulates the reality of Satan during the Middle Ages. So great was his power and the dread that he provoked that he was in many respects a more real, concrete, and insistent reality than God. However, the view of Satan in both the church and the popular culture shifted substantially over time, and in directions that are by no means consistent. Before the thirteenth century, for example, the church did little to encourage belief in the Devil, witchcraft, and other superstitions. Indeed, it frequently tried to obstruct or forbid the prosecution of witches. Yet belief in the devil remained vigorous in popular religion. In effect, the church's efforts to suppress these beliefs were no more successful than the efforts of American elites to suppress nuclear fear. But in the thirteenth century, the church imparted a new impetus to Satan by embarking on the witch craze.[17] Two centuries of organized terror followed. The people were primed for the witchcraft panics by their own vivid and palpable sense of the Devil, as well as by the plagues, famines, and wars that racked the fourteenth and early fifteenth centuries.[18] We focus on the Devil in this period since the witch crazes parallel the nuclear moral panics of the twentieth century.

The Satan of popular Christianity and of folklore cuts a varied and vivid picture. Early medieval thought was spawned in the monasteries, where the monks created a frightful yet still colorful Devil. This version of the diabolical entered public consciousness through homilies, hagiography (stories about the lives and legends of saints), and miracle tales, as well as mystery plays.

Christianity is characterized by an absolute dualism. Satan is the universal personification of evil; in assuming the personal form of the Devil, evil becomes thoroughly demonic. For Satan, who chose himself and not God, aspires to the Godhead. The attempt to usurp the place of the Creator, to be, as Satan was sometimes called, "God of this world," is an unredeemable treachery. Sin exists only when there is inordinate pride. And pride is most prodigal when it mimics God.

Satan was utterly demonic. He was also omnipresent as seducer, tempter, and destroyer. He had been seen by many and assailed everyone. Satan and his followers, the army of lesser devils, demons, enchanters, wizards, and witches, were like a great swarm of flies; the air was so full of demons that a needle dropped from heaven must strike one.[19]

The temptations and assaults that Satan used to test and to pervert people were endless, dramatic, and immoderate. He appeared in

17. See for example, Ben-Yehuda (1980, 1985).
18. Russell (1984, 254). On Satan, see also Graf (1931), Langton (1974), and Russell (1977).
19. Russell (1984, 71).

innumerable guises and employed both the flesh and the inanimate world to conduct his temptations. He could invade the mind and the soul, awakening "riotous thoughts, lawless imaginings, disordered affections, a thousand phantom notions of sin."[20] In sleep, when critical faculties are weakened, he assaults people with visions and dreams that leave behind dangerous agitations and perturbations. Holiness afforded no protection from these affronts. Indeed, his temptations grew more malefic at the most sacred times.

So too was he omnipotent. He had great knowledge, being versed in science, theology, sorcery and magic, sophistry, and literally all other forms of learning. He was a master of crafts, having built many of the bridges, towers, walls, and fortifications in Europe. His powers knew no bounds. Nature did his bidding. He was an untiring worker of misfortunes, employing fraud, guile, and a myriad of other unseemly means to tempt or torture the unwary. Stories of his strength abounded; the enormous boulders dotting the countryside were his work, and there was no shortage of people capable of telling in precise detail how the thing occurred.[21]

Just as the life of each individual was significantly affected by Satan, so too did he influence groups and nations. He instigated heresies, incited conspiracies and rebellions, promoted wars, prepared famines; shipwrecks, murders, and robberies were also frequently his handiwork. Whereas God created "natural plagues," Satan was widely believed to have been responsible for artificial pestilences, which were more virulent than natural ones.

If the Devil appeared when unwanted, he could be summoned as well. Whistling in the dark or writing him a note in Jew's blood and tossing it in fire were among the techniques. Typically he was called up for the purpose of making a formal pact, an idea that extends back to the fifth and sixth centuries.[22] Pact stories, culminating in the legend of Faust, grew in power and detail throughout the Middle Ages. They were not regarded as mere fictions for purposes of instruction: during the witch craze of the fifteenth to seventeenth centuries, they were taken as literal. Central to the hysteria was the notion that witches worshiped Satan and signed a formal pact with him. By the seventeenth century, documents purported to be pacts were presented as evidence in trials of witches and magicians.

During the witch crazes, the people were goaded on by the church. The clergy, and indeed the whole hierarchy, including various popes,

20. Graf (1931, 68).
21. Graf (1931, 57).
22. Russell (1984, 80–91).

vied against each other in the zeal and ferocity with which they carried out God's battle against Satan. By constantly proclaiming the Devil's might and pointing out innumerable examples of his powers over the world; by equating heresy and witchcraft and conducting numerous inquisitions and witch-trials; by expounding the view that hell was much better populated than heaven; and by many other similar acts, the church inflamed the fear of Satan.[23] At times, it may have been more against the Devil than for God. So successful was the effort that by the sixteenth century all magic was popularly attributed to Satan.

The church interwove belief in the Devil with its theology as a means of accommodation with the people. It endeavored to usurp what it could not exterminate. It amalgamated and fermented these formerly suspect ideas in order to achieve its own ends. The witch panics were often inquisitions carried out from above.[24] The people did not initiate them so much as have them impressed on them. They had been well prepared, and fear of the Devil and his witches ran riot.

Still, the church never assumed full command of the popular mind and its myths. Whereas it mobilized Satan's evil as a means of social control, the people, as if to counteract the diabolical, recognized a second, more human Devil. Just as they familiarized Christ, the saints, and God by having them visit the earth, enter homes, sit at tables, and so on, so they ascribed a familiarity to the one who was ubiquitous. Popular folktales depict a domesticated Devil who lives a normal life with normal problems. He is less an object of fear than one of scorn or mirth; conquered or defeated, he is ridiculous and often a target of affection; he has lost so much of his diabolical power that tricks and jokes can be played on him.

The correspondence between the Satan of the witch crazes and the Soviet Union of the nuclear panics is neither perfect nor complete. The closest we come to such a fit is in Hal Lindsey's book, *The Late Great Planet Earth,* where Russia is Gog, the force of absolute evil that will start the next war and capture most of the world just before the Second Coming. The biggest discrepancy, both as Gog and in the more conventional view of the Soviet devil, is the absence of the Soviets as an object of ridicule, as a diabolical power that one can play tricks on. But then nuclear humor has never been big business (pace

23. Graf (1931, 147).
24. Russell (1984, 275).

Dr. Strangelove). As for the Soviets, perhaps the best potential source of humor is their staggering inefficiency and the ironic attitude the Soviet people adopt toward state pronouncements, shortages of (often shoddy) goods, and the interminable lineup.

For the most part, however, there is an uncanny fit between the Soviets and Satan. Up to the Gorbachev revolution (to be discussed in the Conclusion), the postwar world was largely divided into an absolute dualism, pitting two systems, the democratic-capitalist and the totalitarian-socialist, against one another. To be more specific, this divide was especially great between 1949 and 1962, the height of the Cold War. Both détente and the shift toward multipolarity attenuated the bipolar conflict. However, that rivalry was not replaced as the major global issue until the emergence of Gorbachev.

The Soviets are the universal personification of evil. They are thoroughly demonic and aspire to the secular godhead—communist control of the world. Their pride in the certainty of their victory is uncanny. The Soviets are certainly omnipresent, the tempter and destroyer who takes on many guises. On the view that there is a communist under every bed (as well as throughout the State Department), the Soviets are constantly assailing their enemies with conspiracies hatched in the Kremlin. Consider some of the charges routinely leveled against them. They steal military and industrial secrets; they make pacts with spies who do their bidding in the West (Kim Philby et al.); they conduct massive disinformation campaigns, which include the ownership of many newspapers and magazines; they train and arm international terrorists; they have infiltrated and directed the peace movement; they destabilize regimes all over the world; Soviet spies take on the guise of beautiful women to "turn" diplomatic personnel; they cheat on arms-limitation agreements; they are always on the verge of a "breakout" from arms agreements that will provide them with a usable nuclear advantage; they have developed technology to control the weather. While some of these charges can be considered "true," the real issue is the uncritical *exaggeration* of Soviet capabilities—the panic-induced attribution of satanic mastery to a state that was in reality chronically inept.

A typical example of the powers of conspiracy attributed to Moscow is found in a *Reader's Digest* article, "The KGB's Magical War for Peace."[25] According to the article, those opposed to American weapons programs are either communist sympathizers, or, worse, unwitting dupes of Moscow's propaganda. Such is the Soviet ability to

25. Baron (1982).

engineer consent that the nuclear-freeze campaign "thus far has been remarkably successful, for the KGB has induced millions upon millions of honorable, patriotic and sensible people who detest communist tyranny to make common cause with the Soviet Union." Essentially, the freeze campaign was "orchestrated from Moscow" by means of the World Peace Council. Such allegations are not unique. Claims that Moscow organized mass rallies in the streets of New York (on 12 June 1982) were made, for example, in the film *Countdown for America*, narrated by Charlton Heston. Further credence was given these charges by the Committee on the Present Danger, a remarkable organization of conservative American elites, which not only successfully opposed ratification of the SALT II treaty but helped elect Ronald Reagan. As president, Reagan echoed and expanded upon such charges.

Politicians vie with each other in their zeal to be more anticommunist. Political campaigns conducted just on that issue are all too often successful. Richard Nixon early on in his political career learned that he must never let an opponent be more anticommunist than himself. In books and movies, the Soviet villain assumes a commanding role. The greatest challenger in the *Rocky* series was a Russian. Indeed, so great is the strength of this anticommunist sentiment that during the periods of détente with the Soviets American leaders and culture evoked the threat of the "yellow peril."[26] (And when the idea of an ABM system to defend against a massive Soviet attack was made to look ridiculous, some proponents argued that it could be used to defend against a much smaller Chinese nuclear strike.) Effectively, the communist devil was something the United States could hardly do without. It depended on this evil as a rationale for domestic and foreign policy; the conflict provided a sense of national purpose.

Finally, and for us most important of all, Satan was omnipotent, being versed in all knowledge and skills. Such attributes are also ascribed to the Soviets, particularly in their power of subversion, the ability to infiltrate and deceive in order to engineer human behavior. Where Satan used this power to seduce a multitude of shades into his dolorous realm, Soviet might aimed at world domination. When this control and abuse of humanity was coupled with the power of the bomb, the Satan-Soviet connection was fused. In our own terms, they fully realized the potential of totalitarian omnipotence.

Fear of communist subversion preceded the bomb. But in the years

---

26. A colleague, Professor Ralph Beals, suggested this idea to me. I know of no systematic study of the issue.

before the bomb, it never reached the proportions associated with nuclear moral panics. Neither the Red Scare of 1919–20 nor the Great Depression, when Communist party strength in the United States reached an all-time high, evoked the fear and activism associated with McCarthyism. In the prenuclear world, anticommunism focused on "foreigners" or was used to bludgeon nascent working-class movements. The McCarthyist witch-hunts, in contrast, went after targets in the State Department, academia, Hollywood, and the army.

As discussed previously, the postwar hysteria was given much of its initial impetus by atomic-spy stories. These helped foster a mania for secrecy and security. Once the Soviets had the bomb, the restraints that keep political affairs in tolerable bounds were removed. Thus the trial of Julius and Ethel Rosenberg can well be regarded as a "show trial," something usually viewed as a hallmark of totalitarianism. While I ignore the ongoing controversy over the guilt or innocence of the Rosenbergs, it is clear that the trial took place in a context of fear and threat, that it counterpoised good against evil, that all manner of artifice was employed to evoke outrage and alarm, and that there was an urgent need for both scapegoats and assurances in the panic that followed the unexpected explosion of Joe 1. The Rosenbergs were the first convicted spies ever to receive the death sentence from an American civil court. Ethel was the first woman to be sentenced to death by a federal tribunal since Mary Surratt, who was involved in the assassination of President Lincoln.

The feared intersection of the absolute enemy and the absolute weapon was stated succinctly by Churchill in 1948, when he wanted to "bring matters to a head." In his view, if the Soviets were already aggressive while America enjoyed a nuclear monopoly, imagine their behavior "when they got the atomic bomb and have accumulated a large store."[27] Once they had the bomb, all the elements of totalitarian omnipotence were in place. These were stated formally for the first time in NSC-68.[28] Specifically: the Soviets were intent on world domination and hence the destruction of American power; they were absolutely ruthless; their policy was governed only by considerations of expediency; and they would strike when they had sufficient capabilities.

With the Soviet atomic breakthrough, the elements of totalitarian omnipotence were touted and exploited by the MIC. A former head of the Strategic Air Command observed:

27. Cited in Trachtenberg (1988–89, 9).
28. See Sanders (1983) and Trachtenberg (1988–89).

> With such grisly tradition and shocking record in the massacre of their own people, the Soviets cannot be expected to let the risk of even millions of Russian lives deter them from starting a nuclear war if they should consider a war to be in the best interests of the Communist cause. Nor would they be deterred by the danger of losing some cities because widespread devastation and subsequent recovery have numerous precedents in Russia's hectic history.[29]

Albert Wohlstetter, one of the "stars" among American strategic analysts, wrote:

> Russian casualties in World War II were more than 20 million. Yet Russia recovered extremely well from this catastrophe. There are several quite plausible circumstances in the future when the Russians might be quite confident of being able to limit damage to considerably less than this number—if they make sensible choices and we do not. On the other hand, the risk of not striking might at some point appear very great to the Soviets. . . . Then, striking first, by surprise, would be a sensible choice for them, and from their point of view the smaller risk.[30]

These statements could be multiplied indefinitely, for the *ritual* invocation of the Soviet nuclear menace became a near constant in the post-1949 world. Here I discuss the interactions among nuclearism, totalitarian omnipotence, and nuclear forgetting in the world after the Cuban missile crisis.

Nuclearism Revisited

If the impermissibility of war was the first lesson of the missile crisis, the second had to do with strength, particularly nuclear supremacy. Americans were impressed by their own exercise of brinkmanship. A National Security Council Memorandum of 29 October 1962 asserted:

29. Cited in Freedman (1981, 142).
30. Cited in Freedman (1981, 142).

> If we have learned anything from this experience, it is that
> weakness, even only apparent weakness, invites Soviet trans-
> gression. At the same time, firmness in the last analysis will
> force the Soviets to back away from rash initiatives.[31]

Americans were inclined to attribute their "victory" in Cuba to their
superior nuclear forces. Unfortunately, the second lesson presented
in the memorandum was at best half correct. And to make matters
worse, the implications of the two lessons were not combined and
analyzed simultaneously. Where the impermissibility of war aligned
with nuclear forgetting and paralysis, the faith in strength coalesced
with vigilance and the fear of totalitarian omnipotence. The working
out of the paradox of indispensability, in the context of these competing forces that arose from the missile crisis, is what concerns us
here.

Consider first the issue of strength. The missile crisis did reveal
that a preponderance of *conventional* forces conferred a real advantage. Confronted by superior naval and air power in the Cuban
theater, the Soviets had few real options. After the crisis, they expanded their naval and air forces in an effort to match the United
States. In effect, they learned that if they were to be credible militarily, they had to be able to project sea and air power capable of
defending their interests anywhere in the world.

But what about the nuclear lesson? The Russian arsenal, tiny as it
was in comparison to that of the United States, was sufficient to deter
and absolutely petrify the Americans. From the Soviet perspective,
the inferior size and relative vulnerability of their arsenal may well
have been perceived as a disadvantage in the crisis, and they forged
ahead in their effort to achieve parity with the Americans. From the
U.S. perspective, their arsenal was clearly sufficient to deter the
Soviets. But did its larger size confer an advantage on them? I suggest
not. If the arsenal had been large enough and accurate enough to
assure a first-strike capacity—such assurances are nearly impossible,
and the resultant uncertainty is sufficient to deter—then they would
have had a potential advantage. A counterforce strike by the Americans might have stripped the Soviets of their nuclear deterrent (of
course, such a strike that was less than 100 percent effective would
mean thermonuclear attack on American cities). Yet if such a U.S.
advantage existed, it could have been extremely destabilizing. If the
crisis had escalated, the Soviets would have been forced to consider a

31. Cited in Detzer (1979, 260).

preemptive strike, since if they did not use their missiles soon enough they might well lose them. The Americans, aware of this, would then have incentive to fire even earlier.

Such strategic scholasticism notwithstanding, the respective arsenals functioned best to terrify and almost paralyze each side. The crisis was a study in confrontation avoidance, and that is the whole point![32] Because of this scorpion dilemma, adding to the American ICBM force had little strategic significance. Indeed, there were only two strategic reasons for growth in the U.S. arsenal: to prevent the Soviets from attaining a first-strike capacity; and to develop the ability to engage in limited nuclear exchanges, a nuclear war short of an all-out spasm. But given the Soviet inferiority in numbers and technology, the threat of a first strike could hardly be imminent. Moreover, the best assurance against it was to base missiles on submarines. A relatively modest number of these would provide a survivable deterrent. As for limited war, the missile crisis had revealed that the idea of an escalation ladder, so long as there were nuclear rungs, was sheer folly. Yet both of these strategic issues were to become central concerns in the ensuing years.

Where strategic reasons for the indispensability of more weapons were less than compelling, so too was the faith in nuclearism in the post-missile crisis world. Although it upholds the view that the arms race is vital to national security, yet for nuclearism the arms race commands a much broader significance. It comes to encapsulate and symbolize all other conflicts between the forces of good and evil. As a contest, the arms race becomes the ultimate proving ground. Victory is an indicator of superiority, moral and technical. It is a sign of grace, an emblem of American historical transcendence.

But as observed earlier in this chapter, the obsessive fear that had gripped Americans culminated in nuclear forgetting. As a result of the cascading missile panics, nuclearism was now a tarnished faith. The concern with the bomb as a reflection of American civil religion, grace, and moral superiority declined, as one would expect when a

32. Brodie (1973, 426) provides an insightful summary of the missile crisis: "From beginning to end, the confrontation that we call the Cuban missile crisis—the most acute crisis of any we have had since World War II—shows a remarkably different quality from any previous one in history. There is an unprecedented candor, direct personal contact, and at the same time mutual respect between the chief actors. Normal diplomatic formalities of language and circumlocution are disregarded. Both sides at once agree that their quarrel *could* lead to nuclear war, which is impossible to contemplate and which would leave no winner. In effect they were asking each other: How do we get out of this with the absolute minimum of damage to each other including each other's prestige?"

people are trying to obliterate nuclear realities. Fundamentally, Americans no longer wanted to be reminded of—much less confront—the deep vulnerabilities that had been opened by the nuclear genie.

Nuclearism, of course, still persisted, and for several reasons. One had to do with the nature of the weapons themselves. For if the arms race is anything, it is *spectacle*. The source of our most enduring traumas, it evokes visions of vaporized cities, the horrors of radiation poisoning, the terror of sudden death out of a clear sky. It is the source of some of our most searing images, from the ultimate "high" of a nuclear explosion with its flaring mushroom cloud to the *beep—beep—beep* of *Sputnik*. It is the home of strident events, from the trial of the Rosenbergs to the solemn and tense voice of President Kennedy announcing the blockade of Cuba. It is the land of the incredible, where spy satellites read license plates, Atlas rockets ascend in their primeval majesty through a towering ball of fire, and "stealth" bombers are invisible to radar screens. The arms race demarcates new worlds and authors a whole new vocabulary: megatons, megadeaths, megaprojects, megabucks, scenarios, first strikes, Hiroshima, the superbomb, launch on warning, and so on. It seldom ceases to fascinate and to terrify, for the arms race affords endless temptations and opportunities for the Satan of this century. Strategic theorists (and virtually anyone with a grasp of some of the vocabulary) can assume the role of a deity by engaging in the ritual talk that makes the instant death of hundreds of millions possible and acceptable. Nuclear forgetting could numb, but not eradicate, the reality.

Nuclearism also persisted because of the fear of totalitarian omnipotence—Americans could not simply ignore the perceived Soviet threat. The surreptitious attempt to place nuclear missiles in Cuba consolidated the view that the Soviets could not be trusted. Now that the elements of totalitarian omnipotence—world domination, ruthlessness, expediency, and striking when the capacity is sufficient— had been rendered axiomatic, Americans came to see their nuclear superiority as the *only* way of holding the devil at bay. The response to totalitarian omnipotence was founded on two principles: American vigilance and American nuclear strength.

The fear of totalitarian omnipotence appears to have been an authentic response to nuclear moral panics. The MIC did not create these panics so much as impart direction to them, by creating an affinity between national security and the American way of life on the one hand, and nuclear supremacy on the other. That affinity was forged in part by the ritual invocation of threat and of reassurance.[33] The invocation of the Soviet nuclear menace was a near constant in the post-1949 world. From the bomber gap through the "window of vulnerability," from the ABM gap through the MIRV gap and the "throw-weight" gap, Americans were assailed by the acute risks of Soviet perfidy. But in the world after 1962, these gaps did not create moral panics. Rather, they ran headlong into nuclear forgetting. As a result, while the litany of missile counts and comparisons, and the incessant exegeses of a Soviet first-strike potential, occasionally created alarm, they were more likely to be received with apathy, or even to backfire on their proponents.

The ritual invocation of fear was twinned with its opposite, the ritual invocation of reassurance, which was to be found in the development and deployment of the next generation of weapons. While nuclearism remained a hardy faith, reassurance was linked to the idea of a magical weapon that would render American omnipotence "invulnerable."[34] Missiles are not crucifixes or altars; yet, like sacred objects, they exuded a sense of the numinous. Their power seemed unbounded; they were awe inspiring and daunting, for they mediated preservation and death, continuity and finality. Missiles were thus a saving grace. The regeneration they promised required great sacrifices, but the sacrifices conferred a sense of purification and reassurance. They served to stitch and reinforce the tear in the moral canopy.

In the world after the missile crisis, however, belief in invulnerability was rendered far more tenuous. Public attitudes toward the arms race can be better characterized by a fatalistic acquiescence rather than by a sense of pride and awe. To fall behind in the race is intolerable not so much because of the threat to the beliefs embodied in American civil religion as because of the satanic nature of the foe. To not prevail—and here metaphor and reality become hopelessly confounded—is to cede technological omnipotence to the devil. But to relinquish this is to expose decisive deficiencies, perilous shortfalls of national purpose, resolution, and character. All of these must be securely in place, however, if the United States is to withstand the ever-present menace posed by the communist devil. To not prevail, then, is to be bereft of any hope. If the United States cannot retain its technological omnipotence, how is it even imaginable that it can find the ability, resources, and will to resist the myriad of other communist

33. For a technical discussion of ritual in the arms race, see Benford and Kurtz (1987).

34. The term is used by Chernus (1986, 32). On the idea of a magical weapon, see also Boyer (1985, 106) and Benford and Kurtz (1987).

onslaughts and temptations? The United States seeks nuclear supremacy for a host of practical and symbolic reasons under the pretense of
simply defending national security against an implacable and malig-
nant nuclear foe. Chapter 9 examines the ultimate Soviet menace.

9

First Strikes and the Window-of-Vulnerability Panic

Nothing has created so much fear so regularly as the threat of a Soviet first strike. That fear has been a mainstay of totalitarian omnipotence and, by implication, of the arms race itself. The elements of totalitarian omnipotence—world domination, ruthlessness, expediency, and striking when the capacity is sufficient—embody the beliefs necessary to render credible the threat of the Soviets' using their strategic arsenal, either directly or as a means of blackmail. Essentially, the Soviets care only about their ultimate historical destiny: large cities and millions of people can be sacrificed for it. The Soviets are prudent, however, and will not take risks unless the correlation of forces is clearly in their favor. Although American vigilance and nuclear superiority have held them at bay, they are biding their time and planning, waiting for their historic opportunity.

Windows of opportunity that included the possibility of nuclear blackmail or an actual first strike were opened, ostensibly, during the Korean War, in the race for the H-bomb, and in the missile gap that followed *Sputnik*. Ironically, these windows reversed the true state of affairs: it was the United States and not the Soviets who had such a commanding lead that they approached a first-strike potential. To a degree, fear of the Soviets was an externalization or projection of American anxiety about their own numen. The satanic plans attributed to the Soviets had their counterpart in U.S. strategic schemes— the Single Integrated Operational Plans (SIOPs) that, for instance, when the Kennedy administration took office called for an all-out strike against the Soviet Union, China, and most of the East European satellite states. The plan provided for the delivery of 170 atomic and hydrogen bombs on Moscow alone.[1]

At first blush, the ritual invocation of a Soviet first-strike potential amounts to operating, for the sake of security, on the basis of worst-case assumptions. Since Soviet intentions are unknown and totalitarian secrecy renders American intelligence faulty, it is best in the unstable and dangerous nuclear world to assume and plan for the worst. Excessive preparations are inherently more desirable than unpreparedness. Yet as McNamara so aptly put it, worst-case analysis must be balanced by "realistically potential actions."[2] That is just where American analyses miss the mark. Fear of a first strike can be considered the "strategists' conundrum." Every strategic analysis must cut its teeth on that threat. Whatever the strategy, and whether it is proposed by a hawk or a dove, it must treat a Soviet first strike as a veritable threat that it is capable of reducing. For any strategy, nothing matters nearly so much as the capacity to lessen Satan's ultimate temptation. By implication, fear of a first strike is rendered believable by the sheer weight of repetition.

It is in the strategists' conundrum that the Soviet-Satan metaphor acquires its supreme significance. Worst-case assumptions are not based simply on the threat potential inherent in the bomb. Basing deterrence on the logic of assured destruction—this means delivering 400 warheads that would kill 25 percent of the population and destroy a majority of Soviet industrial capacity[3]—is in no way inherent in the confounding power of the bomb. Even the most aggressive states have been deterred historically with far less of a threat. Indeed, a handful of nonoperational and practically nondeliverable atom bombs were ostensibly sufficient to deter the Soviets in the 1940s; planning late in the decade called for a total of 400 atom bombs to deter them. Now each leg of the U.S. strategic triad—ICBMs, bombers, and SLBMs—

1. See Dyer (1985, 250).
2. McNamara (1987).
3. See Myrdal (1977, 118).

can deliver far more than the 400 warheads called for in assured destruction. Still, the fear of a Soviet first strike remains central. Compare this with a statement made by McGeorge Bundy, President Kennedy's special assistant for national security:

> There is an enormous gulf between what political leaders really think about nuclear weapons and what is assumed in complex calculations of relative "advantage" in simulated strategic warfare. Think tank analysts can set levels of "acceptable" damage well up in the tens of millions of lives. They can assume that the loss of dozens of great cities is somehow a real choice for sane men. They are in an unreal world. In the real world of real political leaders—whether here or in the Soviet Union—a decision that would bring even one hydrogen bomb on one city of one's own country would be recognized in advance as a catastrophic blunder; ten bombs on ten cities would be beyond history; and a hundred bombs on a hundred cities are unthinkable.[4]

It is the *intersection* between the bomb and totalitarianism, the mixing of the absolute weapon and the absolute enemy, that produced the totalitarian omnipotence that made utterly unrealistic actions appear feasible. It is as if the United States has been preparing for a struggle against the real Satan, for the potentialities it has attributed to the Soviets are not of this earth. The United States has uncritically generalized from the totalitarianism that Stalin practiced against his own people to the nuclear realm; but the analogy fits poorly.

Whatever the roots and ultimate objectives of Soviet foreign policy, one cannot readily characterize it as *reckless*. Thus it took them several years to consolidate their power in Eastern Europe. In the Iranian crisis following the war, they backed down quickly after an American show of force (one battleship). In Berlin they employed blockades and not direct military force. Perhaps the U.S. bomb restrained them in these situations; however, given the lack of American nuclear preparation in the period of the monopoly, the links remain to be shown. In any case, the Soviets did not become more adventurous once they had their own bomb.

Above all, the Soviets have shown no real recklessness with their nuclear weapons. This is especially evident when their program is

4. Cited in Freedman (1981, 361).

compared to the United States and its nuclear arsenal, where they appear to have been the more cautious of the two powers.

The United States has generally sought to establish and maintain nuclear superiority over the Russians, whether in numbers or in technology. In contrast, the Soviets have generally claimed that they wanted only parity. That they continued to deploy more ICBMs in the 1970s, after the U.S. numbers held steady, has been taken as evidence contradicting this view. That claim is less than compelling. The Soviets built more and larger ICBMs than the Americans to make up for their technical inferiority; miniaturization and electronic advances made U.S. warheads smaller but more accurate and reliable than their Soviet counterparts. At the same time, the Soviets rarely decommissioned aging missiles, with the result that their totals included obsolete liquid-fueled missiles.[5]

The United States has centered its military and defensive strategies around its nuclear arsenal. The Soviets have not relied on their nuclear forces to the same extent. Simultaneously, the United States has settled for what it perceives to be inferiority in conventional forces. Compensation for this disadvantage was to come from "tactical" nuclear weapons. The Soviets, in contrast, have retained larger conventional forces and are presumed to be more prepared to fight a nonnuclear war.

The United States has never rejected first use of nuclear weapons. Indeed, such a declaration would make no sense, given its strategic position. The Soviets did plan for a preemptive strike during the period of their nuclear inferiority. Since they have gained strategic parity in *their* terms, such talk has diminished.

The U.S. concept of deterrence is based largely on the balance of nuclear forces. For the Soviets, deterrence is more a political than a military concept. It entails the idea of avoiding war in the first place. The ability to restrain the opponent is grounded in what they term the "correlation of forces."[6] This involves nuclear and conventional military power, population and economic factors, including civil defense, and less tangible elements as morale and leadership.

"Missile rattling," the overt use of a nuclear signal (such as raising the alert status), has been practiced far more often by the United States than by the Soviets. David Holloway counts nineteen American nuclear "signals," including several in Berlin and the Middle East, as

5. See Holloway (1983).
6. See, for example, Holloway (1983, 82–83).

well as in Korea. In contrast, there is one Soviet incident—the Cuban missile crisis.[7] American missile rattling decreased substantially after 1962.

One cannot make definitive judgments about moral superiority from these considerations. However, they do belie the notion of Soviet recklessness and irresponsibility. The assertion that the USSR has no scruples about using nuclear arms is not supported by the evidence. Of course, the data can be disputed or given various interpretations. Thus, one might maintain that the Soviets do have menacing intentions, but that they have been stymied by either their inferior capabilities or the deterring power of the American arsenal. This analysis would suggest that their public utterances are propaganda used to mask their real objectives.

Looking at selected aspects of Soviet ideology and strategy can lend credence to this view. Until 1953, when Stalin died, official doctrine was based on the "law of the inevitability of war."[8] After his death, the issue was reopened; in 1956, Khrushchev announced that war was no longer "fatalistically inevitable." Propaganda, perhaps; it is also possible that, as a result of nuclear learning, the leadership's strategic thinking was brought more in line with nuclear realities.

While the Soviets declare that they wish to prevent war, they have made extensive preparations to fight one. Given the history of their nuclear vulnerability, coupled with the memory of the Nazi surprise attack in 1941, the Soviets do not want to be caught unprepared. They also plan to prevail in a nuclear war, which can hardly make the United States sanguine. (But then I have seen no evidence that the Americans *plan* to lose such a war! The policy of "escalation dominance" is based on the idea of the United States prevailing at every level of the nuclear-escalation ladder.) Just as significant, however, is the consistent Soviet assertion that a nuclear war would be an utter catastrophe. Preparing to win a war that becomes unavoidable is not the same as having the intention of engaging in war in order to win. The notion of emerging victorious—whatever that means, since "victory" and "survival" have been too often confounded—has definite links with communist ideology. Essentially, it is held that communism will eventually triumph over capitalism, the latter being doomed by the laws of historical development. The time and the process by which

---

7. See Holloway (1983, 51). Holloway's count of Soviet missile rattling can be disputed, since he ignores the threats made during the 1956 Suez crisis. Still, this does not change the basic relationship between the superpowers and nuclear threats.

8. See Holloway (1983, 31).

capitalism will meet its demise is subject to debate and official revision. By implication, a war that destroyed the socialist state would negate the laws of history. To preserve the socialist state and to assure its eventual triumph necessitates either avoiding or prevailing in a nuclear war.

Here we encounter a gap between Soviet rhetoric and Soviet realism. On a political and historical level, the idea of destruction in a nuclear war has been unthinkable. Not to believe in some form of victory would mean that the predicted outcomes of history, on which Soviet ideology and political legitimacy are founded, could be derailed by the technology and the caprice of a historically doomed opponent. In contrast to political assertions, military analyses stress that nuclear war would be an unparalleled catastrophe.[9] There is the clear realization that fighting a nuclear war could not serve any "rational" end of the state. But deterrence could break down. If there were "unassailable, incontrovertible, dire evidence that the United States was about to strike the Soviet Union," the latter would seek to assure the survival of the state through "striking first in the last resort."[10] By contrast, American strategic planning calls for the use of nuclear weapons in the case of a Soviet conventional attack on Europe.

This brings us to the crux of Soviet policy. More than anything else, they have endeavored to establish and preserve the integrity and stability of the state itself. The strength and unity of central power is critical from a pragmatic point of view. It is also an ideological imperative. Socialism cannot rise to its historically ordained heights if the state is sundered in a nuclear exchange. It is this factor, I suggest, that acts as the ultimate restraint on Soviet nuclear intentions. It may well be the case that the leadership of the USSR cares little about human life. Yet much of their murderous madness was in the service of the state. War communism and the various economic plans were an attempt to rapidly and brutally bring the Soviet Union into the twentieth century and compete on an equal footing with capitalism. Domestic opposition, real and imagined, was crushed. The Soviets created their own devil: capitalism and the capitalists, a "mirror image" of the American view. Throughout all this, the Party and the state have remained inviolable.[11]

9. See Erickson (1987).
10. Erickson (1987).
11. The failed coup of August 1991 can be regarded as a last-ditch attempt by former elite communists to preserve their power and the imperial Soviet state. Their actions appear to have been precipitated by the planned signing of the Union Treaty, which aimed to weaken the central power and provide greater independence to the republics.

Writing in *Military Thought* in 1955, Marshal Rotmistrov observed:

> It must be plainly said that when atomic and hydrogen weapons
> are employed . . . surprise attack . . . can cause the rapid
> collapse of a government whose capacity to resist is low as a
> consequence of radical faults in its social and economic struc-
> ture and also as a consequence of an unfavorable geographical
> position.[12]

The only safe conclusion was to prevent a sneak attack.

Given the Soviet response to any form of opposition or dissidence,
it is hardly surprising that there is concern about the stability of the
home front. The problem of nuclear "war" is that it could threaten
this stability, and in ways that are difficult to predict or control.
Certainly most of the Soviet client states in Eastern Europe are
ultimately unreliable. Writing in the context of the possibility of West
Germany acquiring nuclear weapons, Adam Ulam posed the following
problem:

> As to the possibility of a "small" nuclear war, the USSR had to
> think in political terms: against a *small* nuclear power she would
> undoubtedly emerge victorious; but could a *communist regime*
> survive such a war. What would be the consequences of even
> one nuclear missile falling on Moscow and destroying the top
> leadership of the Party and state?[13]

Nuclear restraint, deploying an ABM system around Moscow, and
concern for civil defense: these all serve to reduce the vulnerability
of the Soviet state to nuclear attack. The civil-defense system aims to
protect the top echelons of the Party; the bulk of the population and
industry remain unprotected. Moreover, the effectiveness of these
preparations, to the extent that they would make any difference, is
clearly limited to a small-scale attack. In the same context, the Soviet
effort to stop Chinese nuclear proliferation represented a clear effort
to protect the state against what, in nuclear terms, was a small power.
Similarly, the Soviets have retained tight control over their arsenal
and have not fostered or permitted nuclear proliferation either within
or outside their sphere of influence. The Western record on proliferation is far more dismal.

12. Citation in Holloway (1983, 37).
13. Ulam (1974, 663–64); italics in original.

Technological Gambits and the Window of Vulnerability

From the American perspective, of course, Soviet preparations are simply the prelude to nuclear blackmail or a first strike. In contrast, I am suggesting that the fear of a first strike has more to do with the panic-driven exaggerations of totalitarian omnipotence than with either Soviet capabilities or intentions. In any case, the fear of a first strike has proved to be highly resilient. It was revived again in the late 1970s by the idea of the window of vulnerability, the Soviet potential to destroy 91 percent of U.S. ICBMs. Here I first show how the American desire to exploit a technological lead rendered the window conceivable in the first place. I then examine the emergence and development of the final nuclear moral panics.

In the immediate aftermath of the missile crisis, nuclearism consisted of a blank check for the MIC so that American nuclear superiority could be maintained and any Soviet technical development, no matter how far-fetched, countered. These imperatives were embedded in conventional understandings that were all but unquestioned. However, both Soviet and American actions soon eroded these fundamentals. The first, and by far the most significant, to go was the American superiority in ICBMs. McNamara had halted the U.S. ICBM force at 1,054, and by 1969 the Soviets had caught up.[14] In limiting the arsenal, the defense secretary had realized two things: first, that it would be futile to try to retain a numerical lead; and second, that the American policy of assured destruction had been converted by the Soviet buildup into mutual assured destruction, and there was nothing that could be done about it.

To allay Soviet fears of an American first strike, McNamara publicly downplayed the counterforce option and stressed, instead, assured destruction. (More secretly, the United States retained the capacity for selective counterforce options.)[15] The hope that the Soviets would build about the same number of ICBMs as the Americans proved unfounded. While American numbers remained constant, the Soviets would have about 1,600 ICBMs in 1975.

To put the matter in reverse, because of the futility of trying to retain numerical advantages, the American strategy was to exploit their technological prowess as the only means of maintaining nuclear

14. See Powaski (1987, 114).
15. See Kaplan (1983, 319).

superiority. The irony here is that it was the American attempt to exploit a technological lead that made the window of vulnerability conceivable in the first place. To add paradox to irony, the American developments were fueled by the fear of gaps: somehow, the Soviets, with their inferior technology and far smaller economy, were managing to get ahead! The relevant technologies here are ABMs and multiple warheads, particularly MIRVs.

Between 1955 and 1968, the United States spent more than \$4 billion on ABM plans.[16] But in September 1967, McNamara declared that a nationwide, or "thick," ABM system was of no use, since the expected Soviet buildup would offset its effects. To allay congressional and military pressures, McNamara did announce a "thin" Sentinel system, mostly to protect against a possible Chinese attack. Despite the failure of the American effort, a CIA National Intelligence Estimate suggested that the Soviets might have a nationwide ABM capability by 1966 or 1967. However, construction along the "Tallinn line" proved to be an antiaircraft and not an antimissile system.

While the ABM remained a contested issue among strategists and politicians, efforts to engineer a moral panic over the gap failed. Plans to deploy the ABM created an unanticipated backlash. Where the military had worried that protest would occur in cities not designated as sites, the opposite happened. People worried about "bombs in the backyard," and when the military switched locations of the ABM sites it only created new opponents.[17]

There were a number of reasons for this resistance. First, because of its defensive nature, a Soviet ABM did not directly portend the vaporization of American cities. The threat scenarios for the ABM gap were too theoretical, too remote, and too implausible to engender much terror. As in the case of the failed bomber gap, the threat was not sufficiently spectacular to allow for the construction of a dramatic crisis. Second, the war in Vietnam was now the overriding concern of political activists. Opposition to the ABM meshed with opposition to the war, and the participation of many antiwar scientists resulted in a level of public debate absent from previous nuclear decisions.[18] Third, plans to deploy American ABMs ran headlong into nuclear forgetting, and its corollary, which holds that talk of nuclear war is impermissible. An implication of nuclear forgetting is that the Great Fear can be stoked by *either* Soviet *or* American nuclear developments; anything that suddenly conjures up images of vaporized cities, as in

16. On the ABM, see Chayes and Wiesner (1969) and Kaplan (1983).
17. See Herken (1985, 232–33).
18. See Boyer (1980).

cities that would be targeted by the Soviets because of their ABM sites, seems sufficient to fuel the fear. In effect, the ABM violated the strict concern with deterrence as such, and brought home in selected cities the unacceptable reality of fighting a nuclear war. Such was the resultant opposition to ABMs that when the Nixon administration proposed the Safeguard system to protect Minuteman silos, the authorization passed the Senate by only a single vote in August 1970.

The major impact of ABMs, however, was on the development of multiple warheads.[19] These and other penetration devices were initiated as a response to Soviet attempts to deploy an ABM system; the idea was simply to overwhelm the defenses. (Several other things recommended multiple warheads, including interservice rivalries, industry's need for new weapons systems, and that they increased the "bang for the buck" at a time when the war in Vietnam was limiting funding for new weapons.)[20]

There was concern about Soviet MIRV capabilities, as well. Ritualized warning began as early as 1965. Intelligence reports suggested that they would deploy a system by 1969. While there were indications that such a system was tested that year, it was in all probability a less sophisticated MRV system (multiple reentry vehicles that are not independently targeted). A full-scale MIRV may not have been tested until 1973. Not until the late 1970s did the Soviets deploy highly accurate MIRVs. Here, as with the missile gap and the ABM threat, the USSR was accorded a tremendous lead that had no basis in reality.

The American decision to deploy MIRVs was not taken until mid-1970. Even before then, however, it was clear that MIRVs were dangerous and destabilizing weapons. Since the Soviets deployed more and larger ICBMs to make up for an American lead in the number and accuracy of warheads, when they caught up and MIRVed their ICBMs it would give rise, as was clearly foreseen, to the "window of vulnerability." In particular, the United States worried about the Soviet SS-9, with its large payload capacity. MIRVed, it would present a formidable first-strike threat, since putting multiple warheads on single rockets meant that the warhead-to-target ratio increased dramatically.

Members of the Defense Department and the Arms Control and Disarmament Agency had expressed concern about the counterforce potential inherent in the system. Spokesmen for the latter warned about MIRVs at hearings of the Smith committee. Senator Edward Brooke mounted a concerted campaign to stop their deployment. In

19. On MIRVs, see Greenwood (1975) and Tammen (1973).
20. See Powaski (1987, 115–18).

April of 1969 he met privately with President Nixon and urged him to seek, with the Soviets, an immediate moratorium on MIRV testing and development. The president promised an executive-branch study of the question. Brooke also put similar pressures on members of Congress. Senate Resolution 211, passed in May 1970, highlighted the problems associated with MIRVs and called for a freeze. But deployments began the next month. The main argument that supporters of the program could muster was that it would be very difficult to monitor and verify Soviet compliance with an agreement.

What seems quite clear in this case is that the United States was reluctant to forego a technological lead, one that could temporarily restore its waning sense of omnipotence. Despite panic-ridden gaps, the country had maintained a substantial numerical and technological advantage throughout most of the arms race. By the late 1960s, the USSR was establishing numerical parity. The United States, bogged down in the war in Vietnam, was forced to stand by and watch as its level of strategic superiority slipped and then vanished. In the aftermath of this divisive war, funds for major strategic projects were hard to come by: the blank check had been rescinded. For once other problems, including inflation and long-overlooked domestic needs, took precedence over the communist menace.

Following the signing of SALT I in Moscow in 1972, Nixon, to his chagrin, was forced to more or less recognize parity. The president resurrected the idea of "sufficiency," and both he and Kissinger favored arms limitations rather than another round of technological competition. But not in the case of MIRVs. The United States had a definite advantage here and willingly took the free ride despite the fact that the advantage could only be temporary and would add a new dimension to the nuclear menace.

A further irony was yet to develop. Cognizant of their destabilizing potential, the Nixon administration declared that U.S. policy would be to limit its MIRVs to a strictly retaliatory capacity. In other words, the United States would refrain from deploying the larger warheads and more accurate guidance systems that would give MIRVs the capability to destroy hardened silos. If this was meant to assuage Russian anxieties, it could hardly be very effective. In spite of the limitation placed on MIRVs, the Nixon Doctrine called for enhanced counterforce capabilities—a return to the flexible-response and cities-avoidance strategy of the early Kennedy administration. The administration pushed ahead with the Trident submarine and the B-1 bomber in order to render operational this counterforce option (as well as to create potential "bargaining chips" for SALT).

Air Force Chief of Staff General John Ryan and others would soon speak of the counterforce potential of an improved MIRV system. For the most part, the air force had not given up on the idea of achieving a first-strike potential and continued to seek funds to develop a higher yield and more accurate reentry warhead. The MIRV policy, which could always be changed, thus left the United States ahead, but rendered the lead futile. Without a high level of accuracy, MIRVs served no possible end.

The SALT I talks resulted in the ABM treaty and an agreement to limit strategic offensive weapons. The latter stipulated how many launch vehicles, specifically ICBMs and SLBMs, each side could have. It did not, however, put strict limitations on the qualitative improvement of such vehicles. And it did not limit the number of warheads that could be fixed to the missiles. SALT II did much the same.[21] Hence the SALT process, by way of MIRVs, resulted in a tremendous proliferation of warheads, as both sides took advantage of these "loopholes." The "window of vulnerability," in other words, is a direct consequence of MIRV technology.

The Temptation of Satan

During the 1970s, the Soviets ostensibly developed a new generation of ICBMs capable of attacking missile silos; the United States claims that its own forces did not. The result is a "sanctuary" for Soviet ICBMs and vulnerability for American ones. Thus, calculations for 1982 indicate that the Soviets had the *theoretical* ability to destroy 91 percent of the U.S. ICBMs in a single stroke. This advantage meant that in a nuclear confrontation the Soviets could, as Secretary of Defense Caspar Weinberger told the Senate in 1982, probe "U.S. resolve to retaliate by attacking a smaller and smaller subset of the U.S. military forces—while the U.S. options for retaliation were limited."[22] In other words, a Soviet preemptive strike aimed at Amer-

21. On SALT, see Newhouse (1973); Willrich and Rhinelander (1974); and Wolfe (1979).

22. Caspar Weinberger, "Statement by Secretary of Defense Caspar Weinberger to the Senate," 14 December 1982. Hans Bethe, however, claims that the window is on the other side. On paper, a first-strike using the 550 U.S. Minuteman III MIRVed missiles could destroy more than 90 percent of the modern MIRVed Soviet ICBMs. This would leave the Soviets with mainly obsolete "city-buster" ICBMs, while the United States retained 450 highly accurate Minuteman IIs. The President's Commission on Strategic

ican ICBMs could destroy so many of them that the president would have few options left. To retaliate, he would have to attack Soviet cities, which would mean a riposte against American cities. The results could be paralysis and ultimately surrender to Soviet nuclear blackmail.

Here we have the ritualism of the strategists' conundrum carried to its ludicrous extreme. The scenarios developed by the strategists are logically precise but realistically absurd.[23] For the pyramid of uncertainty that piles up around a first-strike capability is so large and so vexing that even Satan could have serious doubts about such an undertaking. Of course, it is only the notion that one is faced by a satanic foe that affords American strategists the leeway to attribute such intentions and capabilities to the USSR.

A Soviet counterforce strike violates the nuclear firebreak, something both sides have held as sacred, especially since the missile crisis.[24] Once that threshold is crossed, no one knows what will happen. There are no rules of engagement or prior understandings to "control" or "limit" the conflict. The one *theoretical* understanding possibly relevant to a counterforce strike is to avoid cities and thereby minimize collateral damage. But all sorts of factors, ranging from errant missiles to the close proximity of many "military targets" to cities, militate against the viability of this limit.[25] There would also be an unprecedented "fog of war" in a war that is itself unprecedented.

In the first volley, the Soviets would almost certainly destroy the American command, control, and communication system.[26] This would render the tidy and cool conduct of war impossible. There is the added possibility that if communications were cut, the nuclear arsenal would become "fail-deadly"—both submarines and bombers

Forces (the "Scowcroft commission") observed that MIRVed ICBMs offer a favorable "exchange ratio" to the side that fires first. In this regard, the Soviets are more vulnerable since far more of their nuclear power is concentrated in their land-based ICBMs. The Scowcroft commission also concluded that the survivability of the other two legs of the U.S. triad renders the vulnerability of the ICBMs militarily unimportant. See Holdren (1986, 72–76).

23. On strategic theorizing and deterrence, see Catudal (1985); Freedman (1981); Herken (1985); and Jervis (1984).

24. According to Kahn (1965, 26), "no other line of demarcation is at once so clear, so sanctified by convention, so ratified by emotion, so low on the scale of violence, and—perhaps most important of all—so easily defined and understood as the line between not using and using nuclear weapons."

25. Thus the U.S. Defense Department admits that there are up to sixty military targets in Moscow alone. See Powaski (1987, 189).

26. See Ford (1985). Thus Presidential Directive 59, issued by President Carter, calls for the "decapitation" of the Soviet leadership in the first stages of war.

would unleash their deadly cargo. The pyramid will have collapsed on its creators.

Even if the command structure were not destroyed, the very idea of conducting a war without precedent in a cool and rational fashion is fallacious. When McNamara planned the no-cities option, he postulated that the strategic force of each side must "be of a character which will permit its use, in event of attack, in a cool and deliberate fashion and always under the complete control of the constituted authority."[27] Such nuclear cool is like the game of "chicken," where two drivers in the middle of a road drive toward each other at top speed. The one who loses his nerve first and swerves out of the way loses, becoming "chicken," an object of contempt and scorn.

In nuclear chicken, the ostensible aim is a "strategy of rational demolition." There will be rounds of signaling and bargaining, as well as escalation. Each side will engage in a series of tit-for-tat attacks, aimed to destroy the resolve of the other. There will be the need to tolerate the destruction of one of one's own cities for any city one destroys. Both escalating threats and punishments will be used to coerce the other side into an acceptable compromise.

Between "chicken" and the practical strategy it ostensibly authorizes—"nuclear cool"—there lies a radical disjunction, a fissure that admits no bridging. According to Herman Kahn, who introduced the chicken metaphor into the nuclear realm, "chicken" would be a "better analogy" to "cool" if it were played with

> two cars starting an unknown distance apart, travelling toward each other at unknown speeds, and on roads with several forks so that the opposing sides are not certain that they were even on the same road. Both drivers should be giving and receiving threats and promises while they approach each other, and tearful mothers and stern fathers should be lining the sides of the roads urging, respectively, caution and manliness.[28]

The "cool" model assumes that both sides have complete control over their behavior, that both sides will remain completely rational under stress, that both sides only want to "win" and will engage in rational and controlled tit-for-tat exchanges that test the resolve of and seek to coerce the opponent. But none of these assumptions has

27. Cited in Freedman (1981, 232).
28. Kahn (1965, 12).

any basis in crisis behavior. The "assumptions" are really a set of normative injunctions, stipulations that are a poor description of human responses in something as tame as chess and other board games. If chess masters cannot routinely maintain their "cool" (certainly few poker players can), how can both sides do so in nuclear exchanges that have none of the rules, conventions, and traditions of chess?

Crisis behavior is without exception *less* rational than everyday actions. In a crisis, events converge to a high degree of complexity: time pressures increase, as do uncertainties; adequacy of information decreases, as does instrumental control.[29] Decision makers, under extreme stress, are tense, frightened, and soon exhausted, creating problems of judgment and control. But as their judgment becomes poorer, the pressures escalate. They have to sort through contradictory advice, unclear reports about the behavior of the enemy and of their own side, guesses as to what is likely to transpire, and so on. When we toss in the factors unique to nuclear conflict—the lack of any prior experience, the absence of rules of engagement, a particularly murky fog of war, public terror, and the underlying fear of escalation to an all-out exchange—the prospect of a first strike staying limited to "military" targets is indeed remote.

Still other uncertainties can be piled up. The possibility of the Soviets crippling just the American land-based ICBMs is negligible. Even if the 90 percent estimate turned out to be correct, this would still leave about 100 U.S. ICBMs intact. If up to a third of these were MIRVed, this would leave hundreds of warheads from the ICBM force for retaliation, a strike that could be aimed at military targets as well as cities. But all these calculations do not take into account the fact that Russian tests have not, for obvious reasons, been conducted over the parts of the earth where the actual strike would occur. Small differences in geodetic effects would affect the accuracy of the missiles, hence the viability of the strike. Nor can the Soviets test for fratricide, the effects of the interaction between an incoming missile and those that arrive moments later. They would also have to assume that the Americans have not adopted a "launch-on-attack" policy, in which case many of the targeted silos would have been emptied of their death-dealing contents.

Omitted from these calculations are the other two legs of the U.S. strategic triad, the bombers and the SLBMs. Accepting the 90 percent

29. See Kahn (1965, 62).

estimate, the Soviets would destroy 1,960 warheads, only 18 percent of the U.S. total.[30] When all this is put together, we are left with an intriguing question. What would possess the Soviets to see such an attack as reasonable, feasible, and worth the attendant risks? Nothing!—or so it seems. Contemplation of such "silo-busting" scenarios is invariably hypothetical. Soviet behavior is unsituated and unmotivated. Everything is deduced from the theoretical possibilities of absolute weapons in the hands of an absolute evil, not from practicalities and real situations. The Americans are worried about a cosmic risk based on simulations and assumptions that have practically no correspondence with the real world.

In his novel *The World Set Free*, H. G. Wells describes the outbreak of nuclear war as follows: "They went to war in a delirium of panic, in order to use their bombs first."[31] If the superpowers manage to go to war, the leaders will be delirious. But they will not jump to use their weapons first. Indeed, strategic games invariably run into what has been termed the "pattern of political inaction."[32] Such paralysis is also characteristic of real situations. In the Berlin crisis of 1961, the United States held a staggering nuclear advantage that is certainly the closest either side has ever been (and ever will be) to a disarming first-strike capability. Yet U.S. decision makers all said no to a nuclear strike.[33] If the United States was restrained in such an advantageous situation, how is it even conceivable that the Soviets, given the pyramid of uncertainties and the mammoth size of their adversary's arsenal, would fire the hundreds of warheads needed to take advantage of the window of vulnerability?

If the bombs are sent, it will not be because of a failure of deterrence strength. Gorbachev is no more reassured than Bush because his SS-9s can theoretically destroy 91 percent of American ICBMs, while Bush's Minutemen can, according to the worst U.S. figures, destroy only 39 percent of Soviet ICBMs.[34] All the calculations and probabilities will mean only one thing: decision makers will do everything possible *not* to use their arsenals. If they do use them, they will stagger into war, pathetic and defeated victims of their inability to find alternative courses of action.

30. See Holloway (1983, 60–61).
31. Wells (1914, 137).
32. See Jack Beaty, "In Harm's Way," *Atlantic Monthly*, May 1987, 48–49.
33. See Kaplan (1983, 295–301).
34. See Holloway (1983, 60–62). Holloway gives these figures but also disputes them.

The Window of Vulnerability–
Afghanistan Panic

The final panic was, like the first, a mixed one. Its nuclear component was the ostensible "window of vulnerability." The nonnuclear components included the decline of détente, Soviet adventurism in Africa, the Iranian Revolution, and the Soviet invasion of Afghanistan.[35] More than any previous panic, this one involved a concerted propaganda effort by a part of the elite, a revival of the Committee on the Present Danger (CPD).[36] The CPD used the window of vulnerability as the fulcrum of the panic; its position was best stated by Paul Nitze: "To have the advantage at the utmost level of violence helps at every lesser level. In the Korean War, the Berlin blockades, and the Cuban missile crisis the United States had the ultimate edge because of our superiority at the strategic nuclear level. That edge has slipped away."[37] Because of the taken-for-granted affinity among nuclear supremacy, national security, and the capacity to use power coercively, both the failure of détente and Soviet adventurism were directly attributed to their nuclear advantage. In other words, all the international losses by the United States and the international gains by the Soviets were attributed to the ceding of strategic nuclear supremacy.

The loss of the "ultimate edge" became a central issue in the extended debate over SALT II. Critics of the Carter administration claimed that the Soviets had used loopholes in SALT I to achieve nuclear superiority.[38] At the same time, there was also a prolonged debate over détente, as the Soviets continued to violate human rights and support revolutionary movements in the Third World. As a result of these debates and Soviet actions, Carter reversed his initial stance, in which he canceled the B-1 bomber and slowed production of the MX missile, and began a large buildup, which included a NATO agreement to station Pershing and cruise missiles in Europe. Yet the window of vulnerability did not arouse extensive public concern, much less alarm. As an invocation of the Soviet nuclear menace, it

35. An alternative but not necessarily inconsistent view of this panic, which focuses on empire crises, is found in Joseph (1985).

36. See Sanders (1983, 277–93).

37. Cited in Sanders (1983, 256).

38. The cost and unpopularity of the Vietnam War also made it difficult to fund new weapons. Poll results revealing low public support for armaments during the war and the reversal of this attitude in the mid-1970s are discussed by Sanders (1983, 193–94).

was too abstract, too technical, and too implausible to engender a dramatic crisis. It did not entail a new, startling threat evoking images of vaporized cities. Thus the campaign to sell the window would probably have failed if not for other events that galvanized the public: specifically, the Iranian Revolution and the Soviet invasion of Afghanistan. Where it took the fall of China and the Korean War to focus the growing panic over the Soviet atomic bomb, Iran and Afghanistan did the same thing for the window of vulnerability.

The Iranian hostage crisis dramatized, like no previous event, the "Vietnam complex" that had so hindered use of American power. With its "yellow-ribbon" campaign and relentless media coverage, that crisis created an outpouring of emotion and patriotism probably unmatched since World War II. The Soviet invasion of Afghanistan seven weeks after the hostages were seized provided the pretext for a dramatic crisis that was directly attributed to the window of vulnerability.

These crises undermined the vacillating Carter administration (the inept attempt to free the hostages in April 1980 was certainly the low point of the period) and contributed to the Reagan electoral landslide. For the new administration and much of its public, America's losses were a direct result of ceding nuclear supremacy. So central was this issue that of all the elements of Reagan's Soviet policy—anti-Soviet rhetoric, stalling on arms limitations, and the military buildup—only the last remained intact at the end of his presidency. Reagan's panic-induced buildup dwarfed that of the Kennedy/McNamara era and continued even as its adverse economic and social effects became transparent. In his first term, defense spending almost doubled, as the president aimed to restore American nuclear supremacy.[39] He balked at any new arms-control agreements, harped on the theme of Soviet cheating, and threatened to revoke existing agreements, including SALT II (which, although not ratified by the Senate, was being followed in principle) and the ABM treaty.

Reagan also plugged the Carter Doctrine, which called for preparations for a protracted nuclear war and "escalation dominance" (the American ability to coerce the Soviets at every level of the escalation ladder). If these actions were intended to evoke assurance, they backfired. The original panic was not based so much on an authentic surge of nuclear fear as it was an attempt to blame the dramatic crisis encompassing Iran and Afghanistan on the loss of the "ultimate edge." It was these events that allowed the CPD to break through the numbing effect of nuclear forgetting.

39. See Powaski (1987, 187–89).

Ultimately, however, the administration oversold its nuclear solution and created, ironically, an authentic fear of the American nuclear menace. Where strategic theorists have been willing to think the unthinkable and thus make preparations for a "limited nuclear war," this is not the case for the public, particularly in the context of nuclear forgetting. Rather, and consistent with the idea that people prefer to avoid thinking about the threat, the public version of deterrence simply equates strength and security. Reagan's assertion that he had never seen anyone insult Jack Dempsey encapsulates the argument that so long as the U.S. arsenal is at least equal to (but preferably stronger than) the Soviets', national security is intact. The public wanted only vague assurances of American nuclear strength and, except for the extreme Right, simply acquiesced in the seemingly necessary Reagan buildup. Such fatalism did not exist in Europe, however.

When the president went beyond strength and deterrence and entered into the realm of discussing nuclear war, he violated a fundamental taboo: nuclear war—and talk thereof—is simply impermissible. Reagan's policy of globally containing the Soviets with coercive means, his rhetoric about fighting and winning a (protracted) nuclear war, his gaffes (you can call the ICBMs back), his loose references to the Old Testament and to Armageddon, his plans for a civil-defense program that envisioned the evacuation of large American cities before the outbreak of war, and fears of the cruise and Pershing missiles resulting in a war limited to Europe, all served to uncork the Great Fear, first in Europe and then in the United States. These miscalculations reveal just how volatile the faith in and the fear of nuclear weapons is.

Unlike the window of vulnerability, Reagan's inadvertent war-fighting panic unleashed a genuine outburst of nuclear fear. As George Kennan observed, "Never in my thirty-five years of public service have I been so afraid of nuclear war."[40] Linked to and accentuating the Reagan episode was a sudden surge in publicity about nuclear war. In November 1983 the ABC network ran the much-hyped film *The Day After.* Jonathan Schell's book *The Fate of the Earth,* which first appeared in the *New Yorker,* became a major sensation. At around the same time, the idea of "nuclear winter" entered the public vocabulary and provided final proof that nuclear war on any scale was intolerable.[41] Catholic bishops, followed by other religious groups, also began

40. Cited in Van Voorst (1982, 86).
41. See, for example, Ehrlich et al. (1983).

to publicly question the ethics not only of nuclear war but of nuclear preparations themselves. In effect, and perhaps for the first time, the two poles of the numen had separated. The extraordinary fear associated with the bomb was revived, but the corresponding sense of power was not. The Strategic Defense Initiative, discussed in the Conclusion, was largely a failed attempt to resurrect the latter pole. The faith in nuclearism had come undone.

The panic resurrected the peace movement and provided it an unprecedented clout, both because of its size and its international scope. The movement imparted such momentum to the idea of a nuclear freeze that a freeze resolution was defeated in the House of Representatives by only two votes in August 1982. Then, in the summer of 1984, Congress imposed moratoriums on various weapons to assure that the president would negotiate in good faith with the Soviets. In 1985, further increases in the military budget, which had been a standard of the Reagan administration, were terminated. All of this served to focus the pressure of the peace movement and made a strong contribution to the signing of the INF Treaty, which rid Europe of intermediate-range nuclear missiles. The Reagan revolution had come full circle.

Conclusion

Where Jimmy Carter shifted in less than four years as president from advocate of disarmament to arsenal builder, Ronald Reagan underwent the opposite shift. Simultaneously, the public culture was subject to contrary panics, the first Soviet based and the second of American origins. With the subsequent signing of the INF treaty, nuclear issues lost much of their urgency and the public culture retreated to the sanctuary of nuclear forgetting. The remaining talks on arms control are too technical and arcane to engender much interest.

Until the Gulf War, "drift" probably best characterized the nuclear policy of the Bush administration. As the Soviet enemy that provided a rationale for so much of U.S. domestic and foreign policy looked less and less like an enemy, American policy makers seemed to be in a quandary. Gorbachev had outflanked them on nuclear and Cold War issues. Yet the administration made no attempt to seize the initiative. Rather than take a proactive position that might help Gorbachev control the opposition or stay in power, policy makers acted as if they were waiting to see if he could survive. The war in the Persian Gulf, however, seems to have provided a new sense of direction and is associated with a number of potentially significant initiatives by the administration, among them the "new world order," plans to make arms sales more transparent, and the agreement on a strategic arms reduction treaty with the Soviets.

This book has focused on the bomb's confounding power—the interaction between nuclear omnipotence and nuclear fear. Rather than trying to foresee an uncertain future, this chapter examines the possible implications of the forces that we have been tracing. Specifically, I suggest that the extreme emotions associated with *both* poles of the numen have been attenuated. In Chapter 9, I contended that the Reagan war-fighting panic revived the fear, but that for the first time it was decoupled from the faith in omnipotence. Here I first expand on the loss of faith in nuclearism. I then argue that the fear has also diminished, but remains a far more potent factor than the faith. For while there appears to be nothing on the horizon that could resurrect nuclearism, nuclear proliferation could revive the fear.

Omnipotence was ascribed to the bomb from the outset. The affinities among nuclear supremacy, national security, and the American way of life were forged in the Soviet-atomic-bomb and *Sputnik* panics. Soviet challenges to America's cherished beliefs galvanized the emotional energies of the nation and spawned formidable efforts to regain technological preeminence, which was uncritically generalized to historical preeminence.

In the 1970s, however, the beliefs embodied in America's civil religion took a beating. The war in Vietnam, Watergate, the loss of nuclear supremacy, and the Iranian hostage crisis all served to challenge the sense of the country's special destiny; still, its close association with technological supremacy was retained. One indicator was the strength of the reaction to the explosion of the space shuttle *Challenger*. This failure created a wide sense of dismay and renewed the fear that the United States is behind in the space race. In this case, there is the added worry that the country trails not only the Soviets but also the Chinese, Japanese, and Europeans.

This reaction indicates a striking shift in the link between the civil religion and American technological preeminence. Where America's special destiny was once closely tied to its nuclear supremacy, the faith in nuclearism has come undone and is now being replaced by faith in sophisticated conventional military technologies. That nuclear faith has been sundered can be seen in the reaction to Star Wars; that it has shifted to other, nonnuclear technologies can be seen in the reaction to the Gulf War.

Reagan introduced the idea of the Strategic Defense Initiative (SDI) when his other nuclear policies had engendered a moral panic. SDI was presented prematurely, without evidence of its feasibility. It was not the result of lobbying for a space-based defense by the MIC. Rather, it represented a personal fantasy of the president, a dream that drew more nourishment from the faith in America's special destiny than from clear analysis. From the outset, SDI was presented in metaphysical terms. It promised something unique: the restoration of America as the "trustee of history."[1] The arms race had culminated in the utter vulnerability of both sides, with exterminism posing the threat of the "social death." SDI, by promising to liberate humanity from "the prison of mutual terror," was clearly a "Good" technology, one that would guarantee the United States "invulnerable omnipotence." And by creating such a beneficial technology, the country would resume its special moral and historical role. The link between progress and Yankee ingenuity would be forged again.

Speaking to high school students in Fallston, Maryland, after the 1985 Geneva Summit, Reagan linked the various themes:

> I told [Gorbachev] that SDI was a reason to hope, not to fear; that the advance of technology, which originally gave us ballistic missiles, may soon be able to make them obsolete. I told him that with SDI, history had made a positive turn. I told him that men of good will should be rejoicing that our deliverance from the awful threat of nuclear weapons may be on the horizon, and I suggested to him that I saw the hand of Providence in that. What could be more moral than a system based on protecting human life rather than destroying it? I could no more negotiate SDI than I could barter your future.[2]

Reagan is returning here to the theme of a "sacred trust," offering at one point to share this technology with friend and foe. The reference to Providence is important in several respects. It not only provides a possible seal of approval for this Good technology that opposes Bad nuclear technology, but the hand of Providence also provides potential legitimacy for a technological goal that most scientists regard as impossible. Providence additionally provides a context for Reagan to

1. Smith (1987, 19).
2. Cited in Smith (1987, 23).

lecture and lord it over Gorbachev. For by now it was abundantly clear that the Soviets lacked the economic base and technological resources to compete. SDI would impose horrendous costs that would drain them of valuable resources.[3] Here again was a technology that only America could build.

The Strategic Defense Initiative is the apotheosis of nuclearism. From the atom bomb, which promised to be the answer to the Great Collapse, through the moral panics, which authorized the buildups that culminated in exterminism, the United States has wedded its future to omnipotent technologies. However, the idea of an impenetrable defense has been officially dismissed as a misguided fantasy. More important still is that "Star Wars," with all its evocative associations, has had virtually no hold on the public imagination. SDI did not allay the deep fears provoked by the Reagan panic and, like all earlier plans for civil defense or for ABMs, has engendered more indifference or fear than faith or reassurance. In this regard, what is remarkable about SDI is not the opposition that it has engendered, but the extent to which it has gone forward despite such opposition, the high probability that the system cannot work and will be more destabilizing than stabilizing, the improvements in Soviet-American relations, and the massive budget deficits in the United States.

Despite the public loss of faith in nuclearism, planning by political and military elites still relies on American technological might. Consider the report of the Commission on Integrated Long-Term Strategy, which aims to prepare the United States to meet the "security environment" until the early decades of the next century.[4] Following a broad review of the security environment that encompasses changes in global economics, technologies, and politics, the report paradoxically places its faith in the exploitation of American technological superiority. In other words, an acute disparity appears to exist between the broad geopolitical approach at the start of the report and the increasingly narrow concern with "emerging technologies" as the solution to all problems, political, diplomatic, and economic.[5]

The link between conventional technologies and America's destiny has been thrown into relief by the Gulf War. Using the advanced technologies built during the Reagan presidency to fight the Soviets,

3. See Knelman (1985, 24).
4. Iklé and Wohlstetter (1988). Besides the named authors, the report includes inter alios such luminaries as Henry Kissinger and Zbigniew Brzezinski. Given the absence of a minority report, it can be regarded as an overview of American grand strategy as seen by its policy-influencing elite.
5. See Kennedy (1988b).

American air forces convincingly routed their Iraqi counterparts.[6] "Smart" bombs and "stealth" fighters became household terms as television viewers were able to watch weapons home in and destroy their targets. Americans took special pride in the success of the Patriot antimissile missile.

The striking technological feats in the Gulf War have revived claims about America's special historical destiny. In a speech to Congress on 5 March, President Bush said the conflict was the "first test of a new world order coming into view, a world in which there is the very real prospect of a new world order." There is little doubt that the new order, which will require an entente of big powers (or simply Soviet acquiescence), will be led by the United States. On several occasions the president has expressed the view that "the American republic will continue to represent mankind's last, best hope." And in his state-of-the-union speech in January, he proclaimed: "We are Americans. We have a unique responsibility to do the hard work of freedom. And when we do, freedom works." Where just five years ago a major concern was "overreach" and the collapse of the American empire,[7] it is now possible to contend that the coming hundred years will be another American century. Thus the influential journalist Charles Krauthammer has recently suggested in *Foreign Affairs* that the United States use its unique position as the only state capable of projecting economic, military, and diplomatic power around the globe to dictate the values of democracy and the free market to other nations.[8] The Pentagon, which will still face budget reductions, is now talking of a new generation of "brilliant" military technologies. In effect, if the United States cannot compete with Japan in consumer electronics, it does produce superior military technology, which embodies considerable political, diplomatic, and military leverage.[9] All of this, of course, is quite independent of emerging nuclear technologies.

The present analysis has stressed the impact of fear on the arms race. For the first two decades of that race, moral panics served to induce nuclear buildups. But then the cascading fear was partially detached from the idea of omnipotence and served to create backlashes against

6. A report by the Project on Government Procurement claims that the Pentagon exaggerated the effectiveness of high-tech ordnance and that many older, less sophisticated weapons performed better than their new counterparts. See the *Toronto Star*, 8 August 1991, A-19.

7. Kennedy (1988a) is the best known of the overreach theses.

8. Krauthammer (1990–91).

9. Some of the critical computer components for U.S. weapons are produced in Japan. Understandably, this dependence is not publicized.

the arms race. Examples of the latter include the fear of radiation that helped spawn the Partial Test Ban Treaty, protests against the deployment of ABM systems around American cities, and the massive peace movement unleashed by President Reagan's loose nuclear talk that accompanied the deployment of Pershing and cruise missiles in Europe. It was that final outburst of fear that sounded the death knell of nuclearism.

I have also placed the bomb at the center of Soviet-American problems and focused on its impact on the Cold War. Most analyses of the Cold War do not ascribe so much weight to the bomb. Here I suggest that its preeminence is clear in the impact of the Gorbachev revolution on the West. Gorbachev's remarkable entry onto the world stage was authored in the nuclear realm. Other Soviet leaders had already claimed that nuclear war was insane; Gorbachev reiterated the claim, but made it far more believable. What he did, whether intentionally or not, was to effectuate a massive crack in the armor of totalitarian omnipotence. Beyond his obvious charisma—the refreshing candor and the absence of rhetoric that accompanied *glasnost*— Gorbachev's early statements plainly laid out Soviet economic inefficiencies and problems. He made it clear that the only place the Soviets had almost (but not completely) managed to keep up was in the military realm. He further made it clear that this island of productivity could not be sustained, in terms of both developing advanced technologies and the drain on the economy.[10] For a Soviet leader to make such startling revelations—"Kremlinologists" knew about these problems, but their knowledge had little public impact—mattered much more than his claims about *glasnost*. In effect, Gorbachev demonstrated openness in the area where it mattered most: the arms race.

To put the matter another way, admitting economic failure undermined the view that the Soviets were engineers of human souls. That they could neither organize nor motivate economic efficiency, even with all their state controls in place, did serious violence to the stereotype of totalitarian omnipotence. When these confessions were coupled with Gorbachev's personal nuclear diplomacy, culminating in the INF treaty, the harrowing obsession with the bomb was lifted. Where nuclear forgetting involved efforts to suppress or ignore the fear, the *weight* of the fear was now lightened. It was this attenuation

10. In *The Russia House,* John Le Carré (1988) places Soviet inefficiency at the forefront of the Western acceptance of *glasnost*. He parodies official slogans as follows: OUR GREATEST PROGRESS IS IN THE FIELD OF BACKWARDNESS! SOVIET PARALYSIS IS THE MOST PROGRESSIVE IN THE WORLD!

of the nuclear burden, which preceded by at least two years the democratization of communist regimes and the surrender of Soviet control in Eastern Europe, that created the Gorbachev persona and allowed such an easing of tensions that 1988 seemed to create an outbreak of peace. In other words, the Cold War issues followed rather than preceded the nuclear breakthrough and have provided the additional "proof" of Gorbachev's good intentions.

Developments on the Cold War front have further attenuated the idea of totalitarian omnipotence. The widespread rejection of Communist parties by electorates and the rapid emergence of nationalist movements clearly belie the proclaimed Soviet capacity to manipulate and control behavior. For Americans, whose cynical distrust of the Soviets is so deeply ingrained, these revolutionary steps have been necessary to combat the cracked but not shattered fear of totalitarian omnipotence.[11] (In November 1991, as I reread the descriptions of totalitarian omnipotence and the threat of a Soviet first strike in previous chapters, the claims seem not merely erroneous but quaint. Hence I would reiterate that only a panic-induced phenomenon could have made such ideas plausible in the first place! The failed coup of August 1991 is further evidence of the exaggerated fear of communist prowess in the West. Clearly, communist regimes do not have the capacity to engineer social control.)

The Gorbachev revolution has rendered the Soviet-American arms race all but irrelevant. Thus, the idea of a Soviet strike "out of the blue" no longer commands our intellectual or emotional energies.[12] Such an attack has been rendered even more implausible by the signing of the Strategic Arms Reduction Treaty, in which both sides will cut their warheads by around 30 percent. Indeed, the treaty calls

11. Writing just before the democratization of the European communist regimes, Richard Nixon (1988) reiterates virtually all the elements of totalitarian omnipotence. He claims that the "Soviets think in terms of centuries" (35) and that their long-term objective is "global predominance," which they "pursue by all means short of all-out war" (46). He further asserts that "Kremlin leaders are masters at strategic deception, disinformation, subversion and other tactics" (101). He states that Gorbachev does not want nuclear war, but still worries about the American deterrent, the ostensible window of vulnerability, and Soviet nuclear blackmail. "Kremlin leaders put an entirely different value on human life" (76). In addition, "The Soviet system of internal repression is the root cause of its aggressive foreign policy" (54). At several points he claims that "the purpose of the Gorbachev reforms is not to move toward more freedom at home" (45), a prognostication belied by recent developments.

12. For a cogent analysis of why American nuclear policies are incompatible with new political realities, see Bethe, Gottfried, and McNamara (1991). Soviet inefficiencies make a mockery of claims that they could in a single stroke destroy more than 90 percent of U.S. ICBMs.

for the Soviets to destroy half of their most dangerous weapons, the heavy SS-18 missiles. Ultimately, however, the fear of a first strike has vanished because of the collapse of the Soviet state. Ukrainian independence eradicates any realistically possible military threat to the West. In the context of a crumbling union beset by ethnic tensions, fear of the Soviet nuclear arsenal centers on the risk of "unauthorized" use in local conflicts. The threat of "loose nukes" led President Bush, in his 27 September 1991 arms-control address, to announce that his country would unilaterally eliminate land-based tactical nuclear weapons in Europe. The president took this unprecedented step in the hope that the Soviets would follow suit. (So great is the perceived threat here that the Americans have offered to help scrap Soviet tactical nuclear warheads.)

In the United States, these changes have engendered the notion of the "peace dividend." However, given severe budget deficits and the ending of the Cold War, current plans to reduce military spending must be deemed modest. Whereas a 50 percent reduction in the budget is now conceivable, the Pentagon plans to reduce military personnel by 25 percent. Whereas the movement toward minimal or finite deterrence is now feasible, nuclear reductions have been somewhat marginal. Specifically, the unilateral cuts announced by the president do not affect strategic modernization. Funding will continue for the B-2 Stealth bomber, the Strategic Defense Initiative, and the single-warhead Midgetman in silos. Significantly, the administration gave no coherent rationale for these decisions.

Modernization of nuclear (and conventional) forces is being sustained by several factors. As a result of nuclear panics, the MIC is deeply entrenched in the United States. Economically, it has spread its tentacles into every crack and crevice in the nation. Thus, plans to cut the military budget invariably encounter local resistance and require extensive and difficult negotiations. Many corporations, particularly in high-technology areas, are heavily dependent on military contracts. The Gulf War, of course, has helped foster belief in a new generation of high-tech weapons, both because of their glamor and the belief that they can be effective with the smaller fighting forces that are the wave of the future. Politically, Americans want to retain their powerful international military presence. The "new world order" and the revival of the civil religion are both premised on the country's "unipolar" power. Finally, arms agreements and cutbacks have traditionally required the efforts of a concerted peace movement that is able to mobilize public fears. Since the public culture has been enveloped by nuclear forgetting, pressures from the peace movement are—and are likely to remain—very limited.

Domestic opposition to strategic modernization has come from the Congress, which has passed big-ticket programs like the B-2 bomber by tiny majorities. Nuclear funding thus remains in limbo, with its domestic fate contingent on whether the opposition can unite on pressing issues at home—the educational system, health care, a crumbling infrastructure, international competitiveness, and so on.

The New Nuclear Venue

The nuclear arms race could well head for some sort of historical somnolence if it were not for one wild card: the prospect of Third World proliferation.[13] Proliferation is not a new issue, but it has come to the fore with the ending of the Cold War and the conflict in the Persian Gulf. Here one might even conjecture that as the Soviet threat disappeared nuclear fear had to find a new target. Or one could suggest that American leaders tried to engineer a panic over the possibility of an Iraqi bomb.[14] A similar claim could be made about the recent "discovery" of nuclear-weapons programs in North Korea and Iran. To make matters worse, there is the fear of two new sources of proliferation: Soviet nuclear wares could be sold abroad, and there is the possibility of soon-to-be unemployed Soviet nuclear experts selling their skills to the highest bidder.

The venue for the arms race is shifting. So, too, are its dynamics. The venue is now in the Third World, with concentration on a small number of "pariah" states (and terrorist groups) that are believed to be seeking nuclear weapons or other means of mass destruction. That Iraq, to take the best-known example, might have had a bomb is a very alarming prospect in a context where there is little faith in technological countermeasures. The absence of such faith is where the dynamics of the arms race are changing. That is (and unlike the Soviet threat), strengthening the American arsenal is not seen as a means of deterring pariah states. At the same time, there is virtually no faith in a defensive system that can render "nuclear weapons impotent and obsolete."

The Iraqi nuclear effort lends credence to the specter of nuclear

13. On proliferation, see Spector (1987).
14. On the exaggeration of the Iraqi nuclear threat, see Albright and Hibbs (1991).

proliferation among rogue states. The greatest fear, in this case, is that a regional war might go nuclear. There is also concern that small nuclear states in the Third World might attempt nuclear blackmail or use their arsenals to embark on "wars of redistribution."[15] Future events could also create moral panics over potential nuclear threats and thereby give rise to preemptive strikes against suspected targets. The depths of nuclear fear in this context are revealed in a 20 November 1990 CBS News/New York Times poll, which found that while a majority of Americans would not go to war to protect access to Middle East oil, they did support a military effort to prevent Iraq from getting the bomb.[16] The Iraqi nuclear threat has also renewed calls for a ground-based antimissile system to protect the United States against a potential small-power nuclear strike. Finally, the further bombing of Iraqi nuclear facilities was deemed acceptable if that nation did not cooperate fully with United Nations inspection teams.

It has long been a cliché that the nuclear genie cannot be put back in the bottle. Still, one does not want to hear the sound of more corks popping. Changing Soviet-American relations affords a unique opportunity in this respect. Thus far the United States has rejected a comprehensive test ban (CTB). Where previous arguments against a CTB dealt, first, with the problem of "verification" and, second, with the issue of weapon "reliability," recent arguments stress the need for testing to make the warheads "optimally safe."[17] Here I assert that rejection of a CTB is now indefensible, since such a ban is the logical first step in the prevention of nuclear proliferation and will purchase far more safety than further tampering with the electronic control systems and high explosives that go into the construction of nuclear weapons. With the end of the Cold War and the real possibility that the modernization cycles that help sustain the superpower arms race are no longer needed, everything must be done to assure that new venues are not opened up. Hence the need for a comprehensive test ban.

The logic of a CTB does not rest in simply upholding the Nuclear Nonproliferation Treaty or in the superpowers leading by example. Nonproliferation has been quite effective without a CTB, and the pariah states that might "go nuclear" are not likely to be influenced by the nonproliferation treaty or the moral example of the superpowers. Indeed, a CTB does not erase the double standard: a few

15. On Third World threats, see Falk (1983, xviii) and Heilbroner (1974, 43).
16. See Albright and Hibbs (1991, 27).
17. See Hippel (1991).

"declared" states have a host of nuclear weapons, and all others have none. Rather, I deem a CTB necessary to provide the moral authority and political leeway to *enforce* nuclear abstinence in the international arena. The world community should aim to put all nuclear tests beyond the pale.

More specifically, a CTB should be tethered to the U.N. and, using recent U.N. actions against Iraq as a precedent, be subject to international verification and international sanctions. Under the auspices of a CTB, a system of monitoring could be put in place, much of which could be carried out with noninvasive technologies. However, a monitoring system without open inspections could not be 100 percent effective. Thus a country might be able to produce a very small number of unsophisticated atom bombs that could be reliable without testing. However, both H-bombs and warheads to tip missiles are very complex and require testing. Since satellites and other means of verification can identify most tests (especially as the size of the warhead increases), violations would be difficult to conceal.

Since nations that keep their nuclear options open usually do so because of mutual distrust and the attendant fear that a potential enemy could develop a bomb, there must be a very high degree of confidence in the effectiveness of a CTB. Hence I recommend that open inspections be mandatory and subject to a specific regime of sanctions. At the first stage, the regime should require open international inspections as a condition for the transfer of any nuclear materials or technology between countries. At the second stage, any state that interferes with or reneges on such inspections or presents other clear evidence of attempting to violate the test ban should be subject to the same type of sanctions that the international community imposed on Iraq before the Gulf War. In effect, and paralleling the case of South Africa, the international community should cast such states beyond the pale.

At the third stage, if sanctions are not effective, other responses would have to be considered. At the extreme, these would include the possibility of internationally sanctioned military action against the suspected nuclear facilities of the outlaw state. Finally, the promise of international retaliation should be attached to the use or threatened use of nuclear weapons.

The effectiveness of a CTB depends on a number of other contingencies. Presumably, such a ban would end all production of bomb-grade fissionable materials, as well as other bomb components (nuclear triggers, etc.), by the declared nuclear states. Most important, nuclear states should continue to make deep cuts in their arsenals and aim for a strategy of finite or minimal deterrence. In this case, the superpowers would each retain about one thousand (not too accurate) warheads that would be based on invulnerable platforms like submarines or mobile missiles. By implication, nuclear weapons would no longer be used to deter conventional attacks. This is critical, since the vast array of battlefield and other "small" nuclear weapons could then be declared redundant and destroyed. Such an approach would effectively deny pariah states or terrorists the capacity to steal or purchase such weapons in whole or in part.

I have sketched these proposals without any consideration of the numerous difficulties involved. Before the Gulf War and the U.N. decision to undo Iraqi programs to build weapons of mass destruction, I would have deemed these ideas utopian. But the unfolding of events in the Persian Gulf has revealed the extent to which international cooperation is now possible on specific issues. In the aftermath of the Iraqi invasion of Kuwait, U.N. resolutions were realized in weeks. The sanctions imposed were, and remain, highly effective. So too was the international military coalition. Finally, and most significantly, U.N. resolutions after the war have clearly abrogated elements of Iraqi sovereignty.

Whereas national sovereignty was once considered inviolable, it is now under challenge from a host of factors. Environmental threats that include ozone depletion and global warming render the idea of absolute sovereignty nonsense. So too do accidents like Chernobyl. In a highly interdependent world, "ecological security" will eventually take its place beside the more conventional military notion of national security; ultimately, elements of the former will displace elements of the latter.[18] Nuclear war was the first truly global threat, and there has perhaps never been a better opportunity to deal with this menace in a concerted international fashion. The precedent of Iraq should be institutionalized in the U.N. The instruments are in place and the stunning coalition victory in the war may be sufficient to cause future belligerents to pull back in time.

Of the five declared nuclear powers, only the Soviets have given their support to a comprehensive test ban. Resistance by the United States, Britain, France, and China is the main obstacle to a CTB. The question, then, is what could provide the impetus or political will to overcome this long-standing opposition? This brings us back to the fear. Without an acute episode of nuclear fear, these declared powers will not face concerted pressures for either a comprehensive CTB or

18. On ecological security, see Ullman (1983) and Mathews (1989).

other steps to end nuclear proliferation. But if nuclear actions by a pariah state were to create a crisis, this could give rise to precipitous and dangerous responses. In effect, we are still hostage to the terror. Future outbreaks of nuclear fear will probably be necessary to provoke meaningful actions, and one can only hope that the fear will be at the right level to facilitate rational rather than irrational responses.

References

Abel, E.
   1966   *The Missile Crisis*. Philadelphia: Lippincott.
Abolfathi, Farid.
   1980   "Threat, Public Opinion, and Military Spending in the United States, 1930–1980." In P. McGowan and C. Kegley (eds.), *Threats, Weapons, and Foreign Policy*. London: Sage.
Adams, D.
   1987   "Ronald Reagan's Revival: Volunteerism as a Theme in Reagan's Civil Religion." *Sociological Analysis* 48: 17–29.
Albright, David, and Mark Hibbs.
   1991   "Hyping the Iraqi Bomb." *Bulletin of the Atomic Scientists* 47: 26–28.
Aliano, Richard.
   1975   *American Defense Policy from Eisenhower to Kennedy: The Politics of Changing Military Requirements, 1957–1961*. Athens: Ohio University Press.
Allyn, Bruce, James Blight, and David Welch.
   1989–90   "Essence of Revisionism: Moscow, Havana, and the Cuban Missile Crisis." *International Security* 14: 136–72.

Almond, Gabriel, Marvin Chodorow, and Roy Harvey Pearce.
1982   *Progress and Its Discontents*. Berkeley and Los Angeles: University of California Press.
Alperovitz, Gar.
1985   *Atomic Diplomacy: Hiroshima and Potsdam*. New York: Penguin.
Ball, Desmond.
1980   *Politics and Force Levels: The Strategic Missile Program of the Kennedy Administration*. Berkeley and Los Angeles: University of California Press.
Baron, J.
1982   "The KGB's Magical War for 'Peace.'" *Reader's Digest*, October, 205–59.
Bar-Zohar, Michael.
1967   *The Hunt for Atomic Scientists*. London: Arthur Barker.
Baumer, Franklin.
1960   *Religion and the Rise of Skepticism*. New York: Harcourt.
Beard, Edmund.
1976   *Developing the ICBM: A Study in Bureaucratic Politics*. New York: Columbia University Press.
Beaty, Jack.
1987   "In Harm's Way." *Atlantic Monthly*, May, 48–49.
Becker, Carl.
1960   *The Heavenly City of Eighteenth Century Philosophy*. New Haven, Conn.: Yale University Press.
Beckford, James.
1983   "The Restoration of 'Power' to the Sociology of Religion." *Sociological Analysis* 44: 11–32.
Bellah, Robert.
1967   "Civil Religion in America." *Daedalus* 96: 1–21.
1975   *The Broken Covenant: American Civil Religion in a Time of Trial*. New York: Seabury.
1980   "Religion and the Legitimation of the American Republic." In R. Bellah and P. Hammond (eds.), *Varieties of Civil Religion*. San Francisco: Harper and Row.
Benford, Robert, and Lester Kurtz.
1987   "Performing the Nuclear Ceremony: The Arms Race as Ritual." *Journal of Applied Behavioral Sciences* 23: 463–82.
Ben-Yehuda, Nachman.
1980   "The European Witch Craze of the Fourteenth to Seventeenth Centuries: A Sociological Perspective." *American Journal of Sociology* 86: 1–31.
1985   *Deviance and Moral Boundaries: Witchcraft, Occult, Science Fiction, Deviant Scientists, and Scientists*. Chicago: University of Chicago Press.
1986   "The Sociology of Moral Panic: Toward a New Synthesis." *Sociological Quarterly* 27: 495–513.

Bernstein, Jeremy.
  1991   "The Charms of a Physicist." *New York Review of Books*, 11 April, 47–49.
Bethe, Hans, Kurt Gottfried, and Robert McNamara.
  1991   "The Nuclear Threat: A Proposal." *New York Review of Books*, 27 June, 48–50.
Berger, Peter.
  1969   *The Sacred Canopy*. New York: Anchor.
Beschloss, Michael.
  1986   *Mayday: Eisenhower, Khrushchev, and the U-2 Affair*. New York: Harper and Row.
Blackett, P.M.S.
  1949   *Fear, War, and the Bomb: Military and Political Consequences of Atomic Energy*. New York: Whittlesey House.
  1968   "A Check to the Soviet Union." In P. Baker (ed.), *The Atomic Bomb: The Great Decision*. New York: Holt.
Blight, James, Joseph Nye, and David Welch.
  1987   "The Cuban Missile Crisis Revisited." *Foreign Affairs* 66: 170–88.
Bloom, Alexander.
  1986   *Prodigal Sons: The New York Intellectuals and Their World*. New York: Oxford University Press.
Boyer, Paul.
  1980   "From Activism to Apathy: The American People and Nuclear Weapons, 1963–1980." *Journal of American History* 70: 821–44.
  1985   *By the Bomb's Early Light*. New York: Pantheon.
Brodie, Bernard.
  1946   *The Absolute Weapon*. New York: Harcourt.
  1959   *Strategy in the Missile Age*. Princeton, N.J.: Princeton University Press.
  1973   *War and Politics*. New York: Macmillan.
Brown, Seyom.
  1968   *The Faces of Power: Constancy and Change in United States Foreign Policy from Truman to Johnson*. New York: Columbia University Press.
Brune, Lester.
  1985   *The Missile Crisis of October 1962*. Claremont, Calif.: Regina.
Burtt, E. A.
  1955   *The Metaphysical Foundations of Modern Science*. New York: Doubleday.
Bury, J. B.
  1955   *The Idea of Progress: An Inquiry into Its Origins and Growth*. New York: Dover.
Bush, Vannevar.
  1949   *Modern Arms and Free Men: A Discussion of the Role of Science in Preserving Democracy*. New York: Simon and Schuster.
Butterfield, Herbert.
  1957   *The Origins of Modern Science, 1300–1800*. Toronto: Clarke, Irwin.

Camus, Albert.
    1956   *The Rebel: An Essay on Man in Revolt.* Translated by Anthony Bower. New York: Vintage.
Catudal, Honore.
    1985   *Nuclear Deterrence—Does it Deter?* Atlantic Highlands, N.J.: Humanities Press.
Chayes, Abram, and Jerome Wiesner.
    1969   *ABM: An Evaluation of the Decision to Deploy an Antiballistic Missile System.* New York: Harper and Row.
Chernus, Ira.
    1986   *Dr. Strangegod: On the Symbolic Meaning of Nuclear Weapons.* Columbia: University of South Carolina Press.
Churchill, Winston.
    1923   *The World Crisis.* New York: Scribners.
    1948   *The Gathering Storm.* Boston: Houghton Mifflin.
    1953   *Triumph and Tragedy.* Boston: Houghton Mifflin.
Clarfield, Gerard, and William Wiecek.
    1984   *Nuclear America: Military and Civilian Nuclear Power in the United States, 1940–1980.* New York: Harper and Row.
Clark, Ronald.
    1980   *The Greatest Power on Earth: The International Race for Nuclear Supremacy.* New York: Harper and Row.
Cohen, Stanley.
    1972   *Folk Devils and Moral Panics.* London: MacGibbon and Kee.
Cohn, Norman.
    1970   *The Pursuit of the Millennium.* London: Paladin.
Cragg, Gerald.
    1964   *Reason and Authority in the Eighteenth Century.* Cambridge: Cambridge University Press.
Crocker, Lester.
    1956   *An Age of Crisis.* Baltimore, Md.: Johns Hopkins University Press.
Crossman, Richard, ed.
    1950   *The God That Failed.* New York: Bantam.
Detzer, David.
    1979   *The Brink: Cuban Missile Crisis 1962.* New York: Thomas Crowell.
Dingman, Roger.
    1988–89   "Atomic Diplomacy during the Korean War." *International Security* 13: 50–91.
Divine, Robert.
    1962   *The Illusion of Neutrality.* Chicago: University of Chicago Press.
    1971   *The Cuban Missile Crisis.* Chicago: Quadrangle.
    1978   *Blowing in the Wind: The Nuclear Test Ban Debate, 1954–1960.* New York: Oxford University Press.
Donovan, Robert.
    1982   *Tumultuous Years: The Presidency of Harry S. Truman, 1949–1953.* New York: Norton.

Doolittle, T.
1692   *Earthquakes Explained and Practically Improved: Occasioned by the late Earthquake on September 8, 1692 in London.* London: John Saliesbury.
Dunn, Frederick.
1946   "The Common Problem." In B. Brodie (ed.), *The Absolute Weapon.* New York: Harcourt.
Dyer, Gwynne.
1985   *War.* Toronto: Stoddart.
Edelman, Murray.
1971   *Politics as Symbolic Action: Mass Arousal and Quiescence.* New York: Academic Press.
Ehrlich, Paul, Carl Sagan, Donald Kennedy, and Walter Roberts.
1983   *The Cold and the Dark: The World after Nuclear War.* New York: Norton.
Eisenhower, David.
1986   *Eisenhower at War, 1943–1945.* New York: Random House.
Eisenhower, Dwight.
1965   *Waging Peace: The White House Years, 1956–1961.* New York: Doubleday.
Enthoven, Alain, and K. Wayne Smith.
1971   *How Much Is Enough: Shaping the Defense Program, 1961–1969.* New York: Harper and Row.
Erickson, John.
1987   "The Soviet View of Deterrence: A General Survey." In William Evan and Stephen Hilgartner (eds.), *The Arms Race and Nuclear War.* Englewood Cliffs, N.J.: Prentice-Hall.
Etlin, Richard.
1984   *The Architecture of Death: The Transformation of the Cemetery in Eighteenth-Century Paris.* Cambridge, Mass.: MIT Press.
Falk, Richard.
1983   *The End of World Order: Essays on Normative International Relations.* New York: Holmes and Meier.
Feis, Herbert.
1966   *The Atomic Bomb and the End of World War II.* Princeton, N.J.: Princeton University Press.
Fleming, D. F.
1961   *The Cold War and Its Origins, 1917–1959.* New York: Doubleday.
Ford, Daniel.
1985   *The Button: The Pentagon's Command and Control System.* New York: Simon and Schuster.
Freedman, Lawrence.
1981   *The Evolution of Nuclear Strategy.* London: Macmillan.
Fussell, Paul.
1975   *The Great War and Modern Memory.* New York: Oxford University Press.

Galliher, J., and J. Cross.
  1983   *Moral Legislation without Morality*. New Brunswick, N.J.: Rutgers University Press.
Gamson, William.
  1987   "Nuclear Forgetting." *Contemporary Sociology* 16: 15–17.
Garret, W.
  1974   "Troublesome Transcendence: The Supernatural in the Scientific Study of Religion." *Sociological Analysis* 5: 167–80.
Gay, Peter.
  1954   *The Party of Humanity: Essays in the French Enlightenment*. New York: Knopf.
Gehrig, Gail.
  1979   *American Civil Religion: An Assessment*. Norwich, Conn.: Society for the Scientific Study of Religion.
  1981   "The American Civil Religion Debate: A Source for Theory Construction." *Journal for the Scientific Study of Religion* 20: 51–63.
Gilot, Françoise, and Carlton Lake.
  1984   *Life with Picasso*. New York: McGraw-Hill.
Gilpin, Robert.
  1962   *American Scientists and Nuclear Weapons Policy*. Princeton, N.J.: Princeton University Press.
Glacken, Clarence.
  1976   *Traces on the Rhodian Shore: Nature and Culture in Western Thought*. Berkeley and Los Angeles: University of California Press.
Goudsblom, Johan.
  1980   *Nihilism and Culture*. Oxford: Basil Blackwell.
Graf, Arturi.
  1931   *The Story of the Devil*. New York: Macmillan.
Greenwood, Ted.
  1975   *Making the MIRV: A Study in Decision Making*. Cambridge, Mass.: Ballinger.
Groueff, Stephane.
  1967   *Manhattan Project: The Untold Story of the Making of the Atomic Bomb*. Boston: Little, Brown.
Halsell, Grace.
  1986   *Prophecy and Politics: Militant Evangelism on the Road to Nuclear War*. New York: Lawrence Hill.
Hammond, P.
  1980   "The Conditions for Civil Religion: A Comparison of the United States and Mexico." In R. Bellah and P. Hammond (eds.), *Varieties of Civil Religion*. San Francisco: Harper and Row.
Harbutt, Fraser.
  1986   *The Iron Curtain: Churchill, America, and the Origins of the Cold War*. Oxford: Oxford University Press.

Hazard, Paul.
1946   *European Thought in the Eighteenth Century.* New York: World.
1963   *The European Mind, 1680–1715.* London: Pelican.
Heilbroner, Robert.
1974   *An Inquiry into the Human Prospect.* New York: Norton.
Herken, Gregg.
1980   *The Winning Weapon: The Atomic Bomb in the Cold War, 1945–1950.* New York: Knopf.
1985   *Counsels of War.* New York: Knopf.
Hewlett, Richard, and Oscar Anderson.
1962   *The New World.* University Park: Pennsylvania State University Press.
Hilgartner, S., R. Bell, and R. O'Connor.
1982   *Nukespeak: Nuclear Language, Visions, and Mindset.* San Francisco: Sierra Club.
Hippel, Frank Von.
1991   "Warhead Safety." *Bulletin of the Atomic Scientists* 47: 29–31.
Hobsbawm, Eric.
1987   *The Age of Empire, 1875–1914.* London: Weidenfeld and Nicolson.
Hochschild, Arlie.
1979   "Emotion Work, Feeling Rules, and Social Structure." *American Journal of Sociology* 85: 551–75.
Hofstadter, Richard.
1967   *The Paranoid Style in American Politics and Other Essays.* New York: Vintage.
Holdren, John.
1986   "The Dynamics of the Nuclear Arms Race." In Avner Cohen and Steven Lee (eds.), *Nuclear Weapons and the Future of Humanity.* Totowa, N.J.: Rowman and Allanheld.
Holloway, David.
1983   *The Soviet Union and the Arms Race.* New Haven, Conn.: Yale University Press.
Horelick, Arnold, and Myron Rush.
1965   *Strategic Power and Soviet Foreign Policy.* Chicago: University of Chicago Press.
Horowitz, David.
1965   *The Free World Colossus.* London: MacGibbon and Kee.
Howe, Robert.
1978   "Max Weber's Elective Affinities: Sociology within the Bounds of Pure Reason." *American Journal of Sociology* 84: 66–85.
Hughes, H. Stuart.
1958   *Consciousness and Society: The Reorientation of European Social Thought, 1890–1930.* New York: Vintage.
Huntington, Samuel.
1962   *The Common Defense.* New York: Columbia University Press.
Hyde, H.
1980   *The Atom Bomb Spies.* London: Hamish Hamilton.

Iklé, F., and A. Wohlstetter.
    1988   *Discriminate Deterrence: Report of the Commission on Integrated Long-Term Strategy.* Washington, D.C.: U.S. Government Printing Office.
Jervis, Robert.
    1984   *The Illogic of American Nuclear Strategy.* Ithaca, N.Y.: Cornell University Press.
Jonas, Manfred.
    1966   *Isolationism in America, 1935–1941.* Ithaca, N.Y.: Cornell University Press.
Jones, Joseph.
    1955   *The Fifteen Weeks (February 21–June 5, 1947).* New York: Harbinger.
Joseph, Paul.
    1985   "Making Threats: Minimal Deterrence, Extended Deterrence, and Nuclear Warfighting." *Sociological Quarterly* 26: 293–310.
Jungk, Robert.
    1958   *Brighter than a Thousand Suns.* New York: Harcourt, Brace.
Kahn, Herman.
    1965   *On Escalation: Metaphors and Scenarios.* New York: Praeger.
Kaplan, Fred.
    1983   *The Wizards of Armageddon.* New York: Simon and Schuster.
Kaufman, Gordon.
    1985   *Theology for a Nuclear Age.* Manchester: Manchester University Press.
Kaufman, William.
    1964   *The McNamara Strategy.* New York: Harper and Row.
Kendrick, T.
    1956   *The Lisbon Earthquake.* London: Methuen.
Kennan, George.
    1960   *Russia and the West under Lenin and Stalin.* Boston: Little, Brown.
Kennedy, Paul.
    1988a   *The Rise and Fall of the Great Powers: Economic Change and Military Conflict from 1500 to 2000.* London: Fontana.
    1988b   "Not So Grand Strategy." *New York Review of Books,* 12 March, 5–8.
Kennedy, Robert.
    1969   *Thirteen Days: A Memoir of the Cuban Missile Crisis.* New York: Norton.
Kevles, Daniel.
    1978   *The Physicists: The History of a Scientific Community in Modern America.* New York: Knopf.
Keylor, William.
    1984   *The Twentieth-Century World: An International History.* New York: Oxford University Press.
Killian, James.
    1976   *Sputniks, Scientists, and Eisenhower: A Memoir of the First Special Assistant to the President for Science and Technology.* Cambridge, Mass.: MIT Press.

Kindleberger, Charles.
1973   *The World in America, 1929–1939.* Berkeley and Los Angeles: University of California Press.
Knelman, Fred.
1985   *Reagan, God, and the Bomb.* Toronto: McClelland and Stuart.
Koestler, Arthur.
1959   *The Sleepwalkers: A History of Man's Changing Vision of the Universe.* London: Hutchinson.
Koistinen, Paul.
1980   *The Military Industrial Complex: An Historical Perspective.* New York: Praeger.
Krauthammer, Charles.
1990–91   "The Unipolar Movement." *Foreign Affairs* 70: 23–33.
Krell, Gert.
1981   "Capitalism and Armaments: Business Cycles and Defense Spending in the United States, 1945–1979." *Journal of Peace Research* 3: 221–40.
Kundera, Milan.
1984   "The Novel and Europe." *New York Review of Books,* 19 July, 15–19.
Küng, Hans.
1981   *Does God Exist?* Translated by Edward Quinn. New York: Vintage.
Kurtz, Lester.
1988   *The Nuclear Cage: A Sociology of the Arms Race.* Englewood Cliffs, N.J.: Prentice-Hall.
Kurzman, Dan.
1986   *Day of the Bomb: Countdown to Hiroshima.* New York: McGraw-Hill.
LaFeber, W.
1967   *America, Russia, and the Cold War, 1945–1966.* New York: John Wiley.
Lamont, Corliss.
1965   *The Philosophy of Humanism.* New York: Ungar.
Lamont, Lansing.
1965   *Day of Trinity.* New York: Atheneum.
Langton, Edward.
1974   *Satan, a Portrait.* London: Sheppington.
Laqueur, Walter.
1970   *Europe since Hitler.* London: Weidenfeld and Nicolson.
Laurence, William.
1946   *Dawn over Zero.* New York: Knopf.
Le Carré, John.
1988   *The Russia House.* New York: Knopf.
Leiss, William.
1972   *The Domination of Nature.* New York: George Braziller.
Lester, John, Jr.
1968   *Journey through Despair: Transformations in British Literary Culture, 1880–1914.* Princeton, N.J.: Princeton University Press.

Levin, Murray.
  1971   *Political Hysteria in America.* New York: Basic Books.
Lifton, Robert.
  1979   *The Broken Connection: On Death and the Continuity of Life.* New York: Touchstone Books.
  1986   *The Nazi Doctors: Medical Killing and the Psychology of Genocide.* New York: Basic Books.
Lifton, Robert, and Richard Falk.
  1982   *Indefensible Weapons: The Political and Psychological Case against Nuclearism.* Toronto: Canadian Broadcasting Corporation.
Lindsey, Hal.
  1970   *The Late Great Planet Earth.* New York: Bantam.
Liska, George.
  1962   *Nations in Alliance: The Limits of Interdependence.* Baltimore, Md.: Johns Hopkins University Press.
Lo, C.
  1982   "Theories of the State and Business Opposition to Increased Military Spending." *Social Problems* 29: 427–38.
Logsdam, J.
  1970   *The Decision to Go to the Moon: Project Apollo and the National Interest.* Cambridge, Mass.: MIT Press.
Lowen, R.
  1987   "Entering the Atomic Power Race: Science, Industry, and Government." *Political Science Quarterly* 102: 459–79.
Lukács, Georg.
  1963   *The Meaning of Contemporary Realism.* Translated by J. and N. Mander. London: Merlin.
Mailer, Norman.
  1969   *Of a Fire on the Moon.* New York: Grove Press.
Mandelbaum, Michael.
  1981   *The Nuclear Revolution.* Cambridge: Cambridge University Press.
Markoff, J., and D. Regan.
  1981   "The Rise and Fall of Civil Religion: Comparative Perspectives." *Sociological Analysis* 42: 533–52.
Marks, Sally.
  1976   *The Illusion of Peace: Europe's International Relations, 1918–1933.* New York: St. Martin's Press.
Marty, M.
  1974   "Two Kinds of Civil Religion." In R. Richey and D. Jones (eds.), *American Civil Religion.* New York: Harper and Row.
Mathews, Jessica Tuchman.
  1989   "Redefining Security." *Foreign Affairs* 68: 163–77.
McClelland, C.
  1977   "The Anticipation of International Crises: Prospects for Theory and Research." *International Studies Quarterly* 21: 15–38.

McDougall, Walter.
1985   *The Heavens and the Earth: A Political History of the Space Age.* New York: Basic Books.
McNamara, Robert.
1987   *Blundering into Disaster: Surviving the First Century of the Nuclear Age.* New York: Pantheon.
McNeill, William.
1982   *The Pursuit of Power: Technology, Armed Force, and Society since A.D. 1000.* Chicago: University of Chicago Press.
Melman, Seymour.
1970   *Pentagon Capitalism.* New York: McGraw-Hill.
1974   *The Permanent War Economy.* New York: Simon and Schuster.
Mills, C. Wright.
1959   *The Cause of World War Three.* London: Secker and Warburg.
Mintz, Alex, and Alexander Hicks.
1984   "Military Keynesianism in the United States, 1949–1976: Disaggregating Military Expenditures and Their Determination." *American Journal of Sociology* 90: 411–17.
Moskos, Charles.
1973   "The Concept of the Military-Industrial Complex: Radical Critique or Liberal Bogey." *Social Problems* 21: 498–512.
Moss, Norman.
1968   *Men Who Play God.* New York: Harper and Row.
Murray, Robert.
1955   *The Red Scare: A Study in National Hysteria, 1919–20.* Minneapolis: University of Minnesota Press.
Myrdal, Alva.
1977   *The Game of Disarmament: How the United States and Russia Run the Arms Race.* Manchester: Manchester University Press.
Newhouse, John.
1973   *Cold Dawn: The Story of SALT.* New York: Holt.
Nietzsche, Friedrich.
1968   *The Twilight of the Idols.* Translated by R. J. Hollingdale. Harmondsworth: Penguin.
Nisbet, Robert.
1980   *History of the Idea of Progress.* New York: Basic Books.
Nixon, Richard.
1988   *1999: Victory without War.* New York: Simon and Schuster.
Offner, Arnold.
1969   *American Appeasement: United States Foreign Policy and Germany, 1933–1938.* Cambridge, Mass.: Harvard University Press.
Otto, Rudolf.
1958   *The Idea of the Holy.* London: Oxford University Press.
Phillips, Cabell.
1975   *The 1940s: Decade of Triumph and Trouble.* New York: Macmillan.

Powaski, Ronald.
  1987   *March to Armageddon: The United States and the Nuclear Arms Race, 1939 to the Present.* New York: Oxford University Press.
Prados, John.
  1982   *The Soviet Estimate: U.S. Intelligence Analysis and Soviet Strategic Forces.* Princeton, N.J.: Princeton University Press.
Pringle, Peter, and James Spigelman.
  1981   *The Nuclear Barons.* New York: Holt, Rinehart and Winston.
Prins, Gwyn.
  1982   *Defended to Death: A Study of the Nuclear Arms Race.* Harmondsworth: Penguin.
Quester, George.
  1970   *Nuclear Diplomacy: The First Twenty-five Years.* Cambridge, Mass.: Dunellen.
Rhodes, Richard.
  1986   *The Making of the Atomic Bomb.* New York: Simon and Schuster.
Rock, William.
  1977   *British Appeasement in the 1930s.* New York: Norton.
Room, R.
  1976   "Ambivalence as a Sociological Explanation: The Case of Cultural Explanation of Alcohol Abuse." *American Sociological Review* 41: 1047–64.
Roosevelt, Franklin.
  1941   *The Public Papers and Addresses, IX.* New York: Russell and Russell.
Rosenberg, David.
  1979   "American Atomic Strategy and the Hydrogen Bomb Decision." *Journal of American History* 66: 62–87.
  1982   "U.S. Nuclear Stockpile, 1945 to 1950." *Bulletin of the Atomic Scientists* 38: 25–30.
  1983   "The Origins of Overkill: Nuclear Weapons and American Strategy, 1945–1960." *International Security* 7: 3–71.
Rosi, Eugene.
  1964   "How Fifty Periodicals and the *Times* Interpreted the Test Ban Controversy." *Journalism Quarterly* 41: 545–56.
Ross, G.
  1983   *The Great Powers and the Decline of the European State System.* New York: Longman.
Rossi, P.
  1970   *Philosophy, Technology, and the Arts in the Early Modern Era.* New York: Harper and Row.
Russell, Jeffrey Burton.
  1977   *The Devil: Perceptions of Evil from Antiquity to Primitive Christianity.* Ithaca, N.Y.: Cornell University Press.
  1984   *Lucifer: The Devil in the Middle Ages.* Ithaca, N.Y.: Cornell University Press.

Russell, John.
  1981   *The Meaning of Modern Art.* New York: Harper and Row.
Sanders, Jerry.
  1983   *Peddlers of Crisis: The Committee on the Present Danger and the Politics of Containment.* Boston: South End Press.
Schell, Jonathan.
  1982   *The Fate of the Earth.* New York: Knopf.
Schelling, Thomas.
  1966   *Arms and Influence.* New Haven, Conn.: Yale University Press.
Schlesinger, Arthur, Jr.
  1965   *A Thousand Days: John F. Kennedy in the White House.* Boston: Houghton Mifflin.
Schur, Edwin.
  1980   *The Politics of Deviance.* Englewood Cliffs, N.J.: Prentice-Hall.
Shay, Robert.
  1977   *British Rearmament in the Thirties: Politics and Profits.* Princeton, N.J.: Princeton University Press.
Sherwin, Martin.
  1977   *A World Destroyed: The Atomic Bomb and the Grand Alliance.* New York: Vintage.
  1985   "How Well They Meant." *Bulletin of the Atomic Scientists* 41: 9–15.
Sigal, Leon.
  1988   *Fighting to the Finish: Politics of War Termination in the U.S. and Japan, 1945.* Ithaca, N.Y.: Cornell University Press.
Skvorecký, Josef.
  1988   *The Engineer of Human Souls.* Translated by Paul Wilson. Toronto: Lester.
Smith, Alice.
  1965   *A Peril and a Hope: The Scientists' Movement in America, 1945–47.* Chicago: University of Chicago Press.
Smith, Jeff.
  1987   "Reagan, Star Wars, and American Culture." *Bulletin of the Atomic Scientists* 43: 19–25.
Sorensen, Theodore.
  1965   *Kennedy.* New York: Hodder and Stoughton.
Spector, Leonard.
  1987   *Going Nuclear.* Cambridge, Mass.: Ballinger.
Stilgoe, John.
  1983   *Metropolitan Corridor.* New Haven, Conn.: Yale University Press.
Szasz, Ferenc.
  1984   *The Day the Sun Rose Twice.* Albuquerque: University of New Mexico Press.
Tammen, Ronald.
  1973   *MIRV and the Arms Race: An Interpretation of Defense Strategy.* New York: Praeger.

Taylor, A.J.P.
  1964   *The Origins of the Second World War*. Harmondsworth: Penguin.
Thielicke, Helmut.
  1969   *Nihilism: Its Origins and Nature—With a Christian Answer*. New York: Shocken Books.
Thomas, Hugh.
  1987   *The Beginnings of the Cold War, 1945–1946*. New York: Atheneum.
Thompson, Edward.
  1980   "Notes on Exterminism, the Last Stage of Civilization." *New Left Review* 121: 3–32.
Tolstoy, Leo.
  1940   *A Confession, The Gospel in Brief* and *What I Believe*. Translated by Aylmer Maude. London: Oxford University Press.
Trachtenberg, Marc.
  1988–89   "A 'Wasting Asset': American Strategy and the Shifting Nuclear Balance, 1949–1954." *International Security* 13: 5–49.
Turner, James.
  1985   *Without God, without Creed: The Origins of Unbelief in America*. Baltimore, Md.: Johns Hopkins University Press.
Ulam, Adam.
  1974   *Expansion and Coexistence: Soviet Foreign Policy, 1917–1973*. New York: Praeger.
Ullman, Richard.
  1983   "Redefining Security." *International Security* 8: 129–53.
Ungar, Sheldon.
  1990a   "Moral Panics, the Military-Industrial Complex, and the Arms Race." *Sociological Quarterly* 31: 165–86.
  1990b   "Is Nihilism Dead?" *Sociological Analysis* 51: 98–104.
Van Voorst, L. Bruce.
  1982   "The Critical Masses." *Foreign Policy* 46: 82–97.
Vyverberg, Henry.
  1958   *Historical Pessimism in the French Enlightenment*. Cambridge, Mass.: Harvard University Press.
Waddington, P.A.J.
  1986   "Mugging as a Moral Panic: A Question of Proportion." *British Journal of Sociology* 37: 245–59.
Walton, Richard.
  1976   *Henry Wallace, Harry Truman, and the Cold War*. New York: Viking.
Waltz, Kenneth.
  1979   *Theory of International Politics*. Reading, Mass.: Addison-Wesley.
Weart, Spencer.
  1988   *Nuclear Fear: A History of Images*. Cambridge, Mass.: Harvard University Press.
Weber, Max.
  1946   *From Max Weber: Essays in Sociology*. Translated and edited by H. H. Gerth and C. Wright Mills. New York: Oxford University Press.

Wells, H. G.
1914   *The World Set Free.* New York: Dutton.
White, Lynn.
1962   *Medieval Technology and Social Change.* Oxford: Oxford University Press.
Wiley, Basil.
1940   *The Eighteenth Century: Studies in the Idea of Nature in the Thought of the Period.* New York: Columbia University Press.
Willrich, Mason, and John Rhinelander.
1974   *SALT: The Moscow Agreements and Beyond.* New York: Free Press.
Wolfe, Alan.
1984   "Nuclear Fundamentalism Reborn." *World Policy Journal* 12: 87–108.
Wolfe, Thomas.
1979   *The SALT Experience.* Cambridge, Mass.: Ballinger.
Wyden, Peter.
1985   *Day One: Before Hiroshima and After.* New York: Warner.
Yergin, Daniel.
1977   *Shattered Peace: The Origins of the Cold War and the National Security State.* Boston: Houghton Mifflin.
York, Herbert.
1970   *Race to Oblivion.* New York: Simon and Schuster.
1976   *The Advisors: Oppenheimer, Teller, and the Superbomb.* San Francisco: Freeman.

Index